C000016591

John Gamgee

Our Domestic Animals in Health and Disease

John Gamgee

Our Domestic Animals in Health and Disease

ISBN/EAN: 9783348022897

Printed in Europe, USA, Canada, Australia, Japan

Cover: Foto ©berggeist007 / pixelio.de

More available books at **www.hansebooks.com**

OUR
DOMESTIC ANIMALS

IN

HEALTH AND DISEASE

FOURTH DIVISION.—ENZOOTIC DISORDERS—FUNCTIONS
AND DISEASES OF THE NERVOUS SYSTEM—THE
ART OF SHOEING AND LAMENESS.

BY

JOHN GAMGEE

PRINCIPAL OF THE ALBERT VETERINARY COLLEGE, BAYSWATER, LONDON; AUTHOR OF
" DAIRY STOCK ;" " THE VETERINARIAN'S VADE-MECUM," &C. &C.

THIRD THOUSAND

With Numerous Illustrations

[EDINBURGH: MACLACHLAN & STEWART
BOOKSELLERS TO THE UNIVERSITY
LONDON: SIMPKIN, MARSHALL, & CO.

MDCCCLXVII

PRINTED BY NEILL AND COMPANY, EDINBURGH.

CONTENTS OF THE FOURTH DIVISION.

CHAPTER XVI.

ENZOOTIC DISORDERS.

Red water in Sheep.—Sanguineous ascites.—Maladie de sologne.—Pourriture aigue.—Blood disease in lambs.—Navel ill.—Joint ill.—Lamb disease in America.—Parasitic diseases.—General remarks on parasitism.—Parasites never originate spontaneously.—Mode in which entozoa injure and destroy life.—Classification of parasites.—Cystocestoid worms.—Nematoid or round worms.—Trematode or snoking worms.—Measles in the pig.—Measles in cattle.—Hydatids of the liver and other organs.—Echinococcus veterinorum.—Sturdy, gid, or turnsick in cattle and sheep.—Tapeworms in different domestic quadrupeds.—Parasitic lung-disease in lambs, calves, and other animals.—Fluke rot in cattle and sheep.—On pentastoma tænioides of the sheep.—Enzootic diseases of the horse.—Periodic ophthalmia.—Influenza.—Glanders and farcy.

CHAPTER XVII.

NERVOUS ACTION.

Muscular irritability and nervous stimuli.—Reflex actions.—Physical nervous actions.—Mental nervous actions.—Nerve-cells and fibres.—Their functions.—Analysis of nervous matter.—Spinal cord.—Its structure and functions.—Functions of the groups of cells.—Crossing of sensitive impressions on the cord.—The brain.—Oblong medulla.—Pons varolii.—Cerebrum.—Cerebellum.—Their functions.—Uses of the medulla in respiration and deglutition.—Vital point.—Effects of pricking the floor of the fourth ventricle.—Effect of removing the cerebellum.—Turning or rolling as a result of injury to certain parts.—Removal of the cerebrum in birds, and its effects.—Cranial nerves.—Olfactory nerves.—Optic nerves—their functions.—The influence of their crucial and other fibres.—Auditory nerves.—Common motor nerves of the eye.—Pathetic nerve.—Abducent nerves.—Trifacial nerves.—Their varied functions.—Sympathetic ganglia on their course.—Facial nerve—its influence on the salivary glands.—Glosso-pharyngeal nerves—their influence on deglutition.—Pneumogastric nerves.—Chauveau's experiments.—Effects on breathing.—Effects on the lungs when divided.—Action on the heart; on the stomach.—Spinal accessory nerves; their influence on the voice.—Hypoglossal nerves.

CHAPTER XVIII.

DISEASES OF THE NERVOUS SYSTEM.

General remarks.—Rabies canina, or hydrophobia.—Causes.—Geographical distribution; symptoms.—Dumb and barking rabies.—Post-mortem appearances.—Rabies in the horse, ox, sheep, pig, and cat.—Epilepsy; its symptoms.—Dr Brown-Séquard's researches on epilepsy.—Description of convulsions induced artificially.—Treatment of epilepsy.—Catalepsy in the dog; in a wolf; in the horse.—Chorea, or St Vitus's Dance.—Dr Todd's views—symptoms—treatment.—Tetanus—traumatic and idiopathic—symptoms—Nature treatment.—Diseases of the brain.—Vertigo.—Congestion.—Megrims.—Erroneous views commonly entertained on this subject.—Symptoms.—Prevention.—Encephalitis phrenitis, or inflammation of the brain.—Meningitis, or inflammation of the membranous coverings of the brain.—Difficulty of distinguishing inflammation of the brain from that of its meninges.—Causes and symptoms of encephalitis in cattle.—Encephalitis in the horse.—Treatment.—Apoplexy, or extravasation of blood.—Coma.—Immobility, or sleepy staggers.—Softening.—Induration.—Atrophy.—Hypertrophy.—Dropsy, or hydrocephalus.—Tumours.—Diseases of the spinal cord.—Paralysis.—Hemiplegia.—Paraplegia.—Congestion.—Inflammation.—Softening.—Dropsy.—Louping ill in sheep.—Trembling.—Thorter ill.—Cancer of the spinal cord.—Diseases of the nerves.—Neuritis.—Neuroma.

CHAPTER XIX

ON THE FOOT, AND THE ART OF SHOEING.

Horn.—Secreting structures.—Papillæ.—Laminæ.—Horn cells.—Horn fibres.—Growth of human nail and horse's hoof.—On shoeing.—History of the art.—The wall, sole, and frog of a horse's foot.—Bones of the foot.—Peculiarities of the coffin bone.—Prevailing errors on the subject.—Preparation of the horse's foot for shoeing.—The "rogne pied" or toeing knife.—French system of forging shoes.—Art of shoeing in England.—How to make shoes.—Application of machinery.—Form of nails.—The French method of forging shoes.—English shoeing.—Stamping and fullering shoes.—Comparison between English and French nails.—Oriental method of shoeing.—Comparison between English and French shoeing.—Relative labour in making fullered and stamped shoes.—Fullered shoe sometimes specially advantageous.—On the weight of shoes.—Number of nails.—Position of nail-holes.—Toe-pieces.—On fitting the shoe.

CHAPTER XX.

ON THE DISEASES TO WHICH THE FEET OF HORSES ARE SUBJECT.

Contraction of the foot.—Flat or convex soles.—Thrush, its causes and treatment.—Canker.—Corns, their connection with horny tumours.—Sand-crack.—False-quarter.—Fissure.—Keraphylocele.—Seedy toe.—Over-reach.—Treads.—Pricks by nails or other sharp-pointed bodies.—Quittor.—Founder, acute and chronic.—Navicular disease, or navicular joint lameness.

LIST OF ILLUSTRATIONS.

		Vol.	Page
1.	Stomachs of Ruminants, showing the arrangement of the Water-cells in the Camel,	I.	5
2.	Muscles and Nerves on the anterior aspect of the Horse's Head,	I.	7
3.	Horse's Tongue, showing the Papillæ,	I.	9
4.	Ox's Tongue, showing the Papillæ,	I.	1
5.	Parrot Mouth,	I.	11
6.	Fibro-Cartilaginous Pad, with the peculiar Ridges of the hard Palate of the Ox,	I.	13
7.	A Troublesome Bone,	I.	17
8.	Osteoporosis, or Big Head,	I.	20
9.	Bone deprived of its Earthy Phosphates,	I.	30
10.	Temporary Incisor, anterior aspect,	I.	54
11.	Temporary Incisor, longitudinal section,	I.	54
12.	Temporary Incisor, posterior aspect,	I.	54
13.	A Temporary Incisor,	I.	54
14.	Crown and Posterior Aspect of a Temporary Incisor,	I.	54
15.	A Temporary Incisor, indicating wear,	I.	54
16.	A Permanent Incisor,	I.	54
17.	Longitudinal Section of a Permanent Incisor,	I.	54
18.	Crown of Permanent Incisor at different Ages,	I.	54
19.	A simple Tooth,	I.	54
20.	Longitudinal Section of a Simple Tooth, showing the Tooth Pulp,	I.	54
21.	A Molar,	I.	54
22.	Arrangement of Cement on the Crown of a Molar,	I.	54
23.	The Crown of a Molar,	I.	54
24.	The Crown of a Molar,	I.	54

		Vol.	Page
25.	The Form and Position of the Teeth of the Dog,	I.	59
26.	The Milk Teeth in a Puppy two or three weeks old,	I.	60
27.	The Milk Teeth in a year-old Dog,	I.	60
28.	The Teeth of the Dog between two and three years,	I.	60
29.	The Teeth of the Dog between three and four years,	I.	61
30.	The Teeth of the Dog between four and five years,	I.	61
31.	Disease of the Bones of the Temporo-Maxillary Articulation after Open Joint,	I.	69
32.	Strap and Head Collar used in Fractures of the Lower Jaw,	I.	70
33.	Scrofulous Disease of the Lower Jaw,	I.	71
34.	Remarkable condition of the Lower Jaw of the Ox, due to Abscess,	I.	72
35.	Remarkable condition of the Lower Jaw of the Ox, due to Abscess,	I.	73
36.	Fibro-Plastic Degeneration of the Lower Jaw of the Ox,	I.	73
37.	Fibro-Plastic Degeneration of the Upper Jaw of the Ox,	I.	74
38.	Oblique Fracture of the Lower Jaw,	I.	75
39.	Fracture of the Lower Jaw,	I.	76
40.	Consequence of Ununited or Irregularly-joined Fracture of the Lower Jaw,	I.	77
41.	Consequence of Ununited or Irregularly-joined Fracture of the Lower Jaw,	I.	78
42.	Abnormal Position of the Incisors in the Horse,	I.	79
43.	Abnormal Position of the Incisors in the Horse,	I.	80
44.	A Dentinal Tumour,	I.	87
45.	Displacement of a Temporary Molar, with projection of the opposing Tooth,	I.	90
46.	A Balling-Iron,	I.	93
47.	A Balling-Iron,	I.	94
48.	Guarded Tooth-Rasp,	I.	97
49.	The Rabot Odontriteur,	I.	97
50.	The Chiseau Odontriteur,	I.	97
51.	Mr Gowing's Dental Sliding Chisel,	I.	101
52.	Mr Gowing's Guarded Chisel,	I.	101
53.	Gowing's "Lateral Repeller,"	I.	101
54.	Gowing's "Posterior Repeller,"	I.	101
55.	Gowing's Tooth Forceps,	I.	103

		Vol.	Page
56.	Small Tooth Forceps, ...	I.	106
57.	Wendelberg's Simple Tooth-Forceps, . .	I.	106
58.	Pillwax's Tooth-Forceps, ...	I.	106
59.	Gowing's Gum-Lancet, ...	I.	106
60.	Horse's Head, showing the position of the Parotid Glands,	I.	108
61.	The Parotid of the Dog, ...	I.	109
62.	The Sublingual Glands, Ducts, and Openings, .	L.	109
63.	The Sublingual and Submaxillary Glands of the Dog,	L.	110
64.	Section of Horse's Head, showing the Organs of Deglutition therein situated, and their relations, . .	I.	122
65.	The Œsophagean Canal in Ruminants, . .	I.	124
66.	The Muscular Coat of the Œsophagus, . .	I.	125
67.	The Stomachs of Ruminants, . . .	L.	130
68.	The Stomachs of Ruminants, showing the Water Bags of the Camel, ...	I.	132
69.	Section of the Paunch and Reticulum, . .	I.	133
70.	Colin's experiment with Sutures in the Œsophagean Pillars, ...	I.	138
71.	The Œsophagean Pillar of the Lama and Dromedary,	I.	139
72.	The Movements of the Food in the Paunch, . .	I.	140
73.	The Stomach of Monogastric Animals, . .	L.	143
74.	The Stomach of the Pig, ...	L.	144
75.	The Crop of a Pigeon, ...	I.	146
76.	The Alimentary Canal of the Domestic Fowls, .	I.	148
77.	Imaginary Spiral Valve at the Cardiac Opening of the Horse's Stomach, ...	I.	151
78.	Dilatation of the Lower End of the Œsophagus, .	I.	152
79.	A Thorn producing Choking, . . .	I.	159
80.	Forceps for Choking in Cattle, . . .	I.	163
81.	Impaction of the Crop, ...	L.	195
82.	Stomach of the Horse, ...	I.	206
83.	Free Surface of Gastric Mucous Membrane, viewed from above, from Pig's Stomach, cardiac portions, magnified 70 diameters, ...	I.	207
84.	Free Surface of Gastric Mucous Membrane, viewed in vertical section from Pig's Stomach, pyloric portion, magnified 420 diameters, ...	I.	207
85.	Mucous Membrane of Pig's Stomach, from pyloric por-		

		Vol.	Page
	tion, vertical section showing Gastric Tubules and a closed Follicle, magnified 70 diameters, . .	I.	208
86.	Gastric Tubules from Pig's Stomach, pyloric portion, showing their Cœcal Extremities. A Cylindrical Cast of epithelium pressed out from a tubule, showing its size, ...	I.	208
87.	Silver Tube for Gastric Fistula, . . .	I.	210
88.	Position of the Tube in the Stomach in operations for Gastric Fistulæ, ...	I.	211
89.	Union of the Coats of the Stomach, with the Abdominal Walls in forming Gastric Fistulæ, . .	I.	202
90.	Portion of Dog's Intestine, showing the Glands of Brunner, enlarged five times, ...	I.	229
91.	The Large Intestines, ...	I.	231
92.	The Intestines of Ruminants, . . .	I.	235
93.	The Omentum, ...	I.	236
94.	The Liver of the Ox, ...	I.	238
95.	Pancreas, and Pancreatic Ducts in the Dog, .	I.	244
96.	Pancreas, and Duodenum of the Cat, .	I.	245
97.	Supplementary Pancreatic Glands in the Ox, .	I.	246
98.	Biliary Duct, opened to show the small orifices which constitute the apertures of the secondary Pancreatic Ducts, ...	I.	247
99.	Method of collecting the Pancreatic Secretion, .	I.	248
100.	Oathair Calculus, with a phosphatic one imbedded in it, ...	I.	256
101.	Enema Funnel, ...	I.	268
102.	Invagination of the Cœcum, . . .	I.	277
103.	Strangulation of the Intestines, by a pedunculated tumour, ...	I.	278
104.	Strangulation of the Intestines, by a hypertrophied epiploic appendix, ...	I.	280
105.	Inguinal Hernia, ...	I.	306
106.	Scrotal Hernia, ...	I.	308
107.	Deposit from the inside of the Gall Ducts, . .	I.	319
108.	Tesselated Epithelium from Pericardium of Sheep. Fusiform cells are shown, such as are constantly seen in young animals	I.	323

Vol. Page·

109. The Heart and principal Blood-vessels seen from the left side, ... I.   323

110. The Heart and its principal Vessels seen from the right side, ... I.   324

111. Bone of the Ox's heart, natural size, . . I.   325

112. Elastic Network from the tunica media of the Pulmonary Artery of the Horse, with holes in the fibres. Magnified 350 times, ... I.   330

113. Elastic Membrane, from tunica media of the carotid of the Horse. Magnified 350 times, . . I.   330

114. Diagram of the Circulation, . . . I.   338

115. Cardiac Polypus, ... I.   377

116. Cardiac Polypus, ... I.   380

117. Vascular Tumour in Heart of Horse, . . I.   383

118. Thickening of the Endocardium, as the result of inflammation, ... I.   395

119. Phlebolite, ... I.   443

120. Phlebolite, ... I.   443

121. Phlebolite, ... I.   443

122. Phlebolite, ... I.   443

123. Antero-Posterior Section of the Head, showing the entire Mouth, Pharynx, and Nasal Cavities, . I.   451

124. The Vomer . . . . . I.   452

125. Trachea, Bronchia, and Lungs (superior aspect), . I.   456

126. Ultimate Bronchial Tubes, with air sacs connected with it (of a Cat), ... I.   458

127. A very Thin Slice of Cat's Lung (injected, inflated, and dried), from the surface of the lung. Represents the cut aspect, ... I.   460

128. A Thin Slice of a Cat's Lung, . . . I.   460

129. Drawing of a thin slice cut transversely of the Lung of Cat. It shows the shape of the air-sacs, . I.   461

130. Theoretical View of the Pulmonary Tissue, viz., terminal bronchial tube, its dilated extremity, and its groups of air-sacs, divided transversely, . . . I.   461

131. Ramifications of the Sustinent Arteries on the external surface of the Bronchial Tubes, . . I.   465

132. View of the Horse's Chest, indicating the position of

		VoL	Page
	the ribs, and extent of the thorax over which auscultation is performed,	I.	510
133.	Horse's Chest opened from the right side, and indicating the positions of the Heart and Lungs,	I.	511
134.	Cow, showing the extent of surface for auscultation on the right side,	I.	514
135.	View of Left Lung of Ox *in situ,*	I.	515
136.	Stomach Pump,	I.	544
137.	Professor Rey's Tube,	I.	545
138.	Section of Horse's Head, showing the cranial cavity, nasal chambers, turbinated bones, and opening of the Antrum maxillare,	I.	547
139.	Iron Staff, for hyovertibrotomy,	I.	551
140.	Gunther's Instrument for Washing-out the Guttural pouch,	I.	551
141.	Portrait of a Cart Horse with Abscess of the superior Turbinated Bone,	I.	553
142.	Drawing of Disease of the Nose due to a projecting Molar,	I.	554
143.	Nasal Polypus,	I.	621
144.	Closure of the Right Nasal Chamber by a Fold of Mucous Membrane,	I.	623
145.	Showing the Atrophy of the Laryngeal Muscles in *roaring,*	I.	625
146.	Distortion of the Trachea, producing roaring,	I.	626
147.	Deposit in the Trachea, producing roaring,	I.	627
148.	The Lobulated External Surface of the Kidney of the Ox,	II.	3
149.	The Kidney of the Horse,	II.	4
150.	Section of Kidney of Sheep,	II.	4
151.	Malpighian Body and Tube of Ferrein,	II.	5
152.	Malpighian Body, &c. from the Horse,	II.	5
153.	Plan of Renal Circulation,	II.	6
154.	Crystals of Pure Urea,	II.	10
155	Crystals of Nitrate of Urea	II.	11
156.	Crystals of Oxalate of Urea, obtained by adding oxalic acid to concentrated urine,	II.	12
157.	Crystals of Uric Acid,	II.	13

		Vol.	Page
158.	Crystals of Hippuric Acid, ...	II.	15
159.	Crystals of Triple Phosphate, showing their form, .	II.	18
160.	Crystals of Carbonate of Lime, seen by reflected light,	II.	19
161.	Urinometer or Urogravimeter, ...	II.	21
162.	Professor Brogniez's Catheter, ...	II.	37
163.	Instrument for dilating the Urethra, . . .	II.	38
164.	The Penis of the Ox, ...	II.	38
165.	Section of Horse's Skin, enlarged 30 diameters, .	II.	68
166.	Section of Sheep's Skin, ...	II.	70
167.	Skin of Dog, showing the hair-follicles and hairs, .	II.	70
168.	Sweat Glands of the Ox, showing the simplicity of their forms, ...	II.	70
169.	Simple Oil-Glands from near the foot of the Sheep,	II.	72
170.	Compound Oil-Glands from the skin of the mammæ in the Ewe, ...	II.	72
171.	Stem of Hair, Hair Bulb and Follicle, and Sebaceous Gland, ...	II.	72
172.	Compound Sebaceous Gland, from skin of Dog's scrotum, &c.	II.	73
173.	Sebaceous Gland from the skin of Horse's scrotum, &c.	II.	73
174.	Showing the arrangements of the Glands and Wool,	II.	74
175.	Hair-Follicle from the interdigital canal of the Sheep, showing the oil-glands, ...	II.	74
176.	A Hair-Papilla, showing its relations with the hair-bulb and lining of the hair-follicle, . . .	II.	78
177.	Portion of Stem of Wool, showing the imbricated scales of the cortical substance, ...	II.	78
178.	Transverse Section of a Developing Hair in its Follicle,	II.	78
179.	Transverse Section of the Inner Sheath of the Hair in the Pig, ...	II.	78
180.	Showing the Muscles connected with Hairs, .	II.	83
181.	A Sheep Bath, ...	II.	92
182.	Construction of the Roman Bath, . . .	II.	100
183.	Elephantiasis, ...	II.	142
184.	Drawing of Horse affected with *Herpes circinatus*, or Vesicular Ringworm, ...	II.	148
185.	Sarcoptes equi, ...	II.	165
186.	Dermatodectes equi, ...	II.	168

		Vol.	Page
187.	Symbiotes equi,	II.	169
188.	Dermatodectes bovis,	II.	172
189.	Symbiotes bovis,	II.	173
190.	Dermatodectes ovis,	II.	174
191.	Symbiotes suis,	II.	175
192.	Sarcoptes canis,	II.	176
193.	Sarcoptes cati,	II.	177
194.	Louse of Ox (*Hæmatopinus Eurysternus*)	II.	197
195.	Louse of Calf (*Hæmatopinus vituli*)	II.	197
196.	Louse of Ass (*Hæmatopinus asini*),	II.	197
197.	Louse of Swine (*Hæmatopinus suis*),	II.	197
198.	Louse of Dog (*Hæmatopinus piliferus*),	II.	197
199.	Louse of the Horse (*Trichodectes equi*),	II.	197
200.	Louse of the Ox (*Trichodectes scalaris*),	II.	197
201.	Louse of the Sheep (*Trichodectes sphærocephalus*),	II.	197
202.	Louse of the Dog (*Trichodectes latus*),	II.	197
203.	*Elephantiasis Bovina,*	II.	224
204.	Cow-pox (*Variola vaccinæ*),	II.	237
205.	Teat Siphon,	II.	238
206.	The Papular Stage of Smallpox in Sheep,	II.	252
207.	The Pustular Stage of Smallpox in Sheep,	II.	253
208.	Multipolar Nerve-Cell, showing its nucleus and nucleolus,	II.	389
209.	Tubular or White Nerve-Fibres,	II.	390
210.	Portion of Spinal Cord, showing its cut end,	II.	397
211.	Transverse Section of the Spinal Cord and its Membranes,	II.	398
212.	Transverse Section through the Anterior Horn and entrance of a Nerve-root,	II.	401
213.	Diagram showing the crossing of the Sensory Fibres in the Spinal Cord,	II.	408
214.	The Lower Part of the Encephalon,	II.	414
215.	Transverse Section of the Cerebrum, showing the lateral Ventricles,	II.	415
216.	Lateral and Inferior view of Convex Soles,	II.	580
217.	Horny Tumour of the Laminæ and Coffin-bone showing the amount of Absorption.	II.	596

CHAPTER XVI.

ENZOOTIC DISORDERS.

Red water in sheep.—Sanguineous ascites.—Maladie de sologne.—Pourriture aigue.—Blood disease in lambs.—Navel ill.—Joint ill.—Lamb disease in America.—Parasitic diseases.—General remarks on parasitism.—Parasites never originate spontaneously.—Mode in which entozoa injure and destroy life.—Classification of parasites.—Cystocestoid worms.—Nematoid or round worms.—Trematode or sucking worms.—Measles in the pig.—Measles in cattle.—Hydatids of the liver and other organs.—Echinococcus veterinorum.—Sturdy, gid, or turnsick in cattle and sheep.—Tapeworms in different domestic quadrupeds.—Parasitic lung disease in lambs, calves, and other animals.—Fluke rot in cattle and sheep.—On pentastoma tænioides of the sheep.—Enzootic diseases of the horse.—Periodic ophthalmia.—Influenza.—Glanders and farcy.

Red Water in Sheep.—Sanguineous Ascites.— Maladie de Sologne.—Maladie Rouge. —Pourriture Aigue.

THIS enzootic disease prevails to a considerable extent in Ireland. It appears also to be a disease of modern date, and the first and best account of it in this country, was published by Mr T. W. Gowing, a highly intelligent veterinarian, residing in Camden Town, London. Mr Gowing's observations were on a farm in Middlesex, and the disease appeared on rich ground, where the grass was good, and no noxious plants existed on it. During the same year—1849—and since, many outbreaks have been alluded to, and the mortality induced by the disease is occasionally very great.

From the importance of the subject, I am induced to extract the following from Mr Gowing's report:—

"The disease in question was first noticed by the person who had charge of the sheep, in the month of April; he says:—"On visiting them early one morning, I found two lambs very unwell, they staggered in their gait, had separated themselves from the others, were dull and dispirited, their heads drooping, their mouths closed so firmly that I could with great difficulty only open the jaws, and a frothy saliva covered their lips. I administered to each a small quantity of castor oil, mixed with some warm milk, and as they appeared a few hours afterwards to be somewhat better, I placed them again with the ewes. The medicine having operated, they gradually recovered; but remained very weak for several days. From this time all went on well, until the commencement of the following month (May). I had left the flock apparently in perfect health over-night, but on the succeeding morning, one of the ewes presented similar symptoms to the lambs; in addition, however, she was considerably hoven, and breathed with much difficulty. I immediately gave her a full dose of castor oil, and had her walked about very slowly; this, however, caused evident distress. As no fæces had passed by noon, I repeated the oil, and late in the afternoon, the ewe being still in the same state, I determined on exhibiting a saline aperient, but she died while taking it. On opening the body, the paunch burst from the pressure of the great mass of food which it contained; and as little or no gas escaped, I concluded she had overgorged herself, which produced a stoppage in the bowels and death.

" 'A few days afterwards, a lamb, which to all appearance was well at noon, was found dead about 4 P.M.; and on the third succeeding day another was discovered dead, on the flock being visited early in the morning. A third was taken ill two days after this, to which I gave a dose of castor oil, and bled it in the eye and ear veins; the blood was very dark in colour, and flowed slowly. This lamb lived until the following morning: the breathing was laboured and difficult, and at intervals was suspended for several seconds. The medicine operated freely, but no diminution in the severity of the symptoms was observable.'

" The lamb above alluded to was sent to my establishment, and my attendance on the flock was also requested, in consequence of the serious and fatal character the disease had now assumed. In conjunction with my friend, Mr Varnell, I instituted a *post-mortem* examination of the lamb, and found the following lesions:—

" The abdominal viscera were free from structural disease; but the chylopoietic veins generally were distended with dark blood. The biliary ducts and gall-bladder were also very full of bile. The liver was larger than natural, and darker in colour than we usually find it; the spleen normal; the lungs slightly congested, and a small quantity of limpid fluid in each pleural sac; the thymus gland large and dark in colour, which seemingly depended upon venous congestion; the pericardial sac contained about two ounces of fibrin and serum—the fibrin was in a state of semi-coagulation, but not adhering to any part of the membrane, which showed no redness or abnormal thickening. The external part of the heart, particularly on the left side, was observed to be studded with dark-looking spots.

" On making a section through the outer wall of the right ventricle, from its base to its apex, the cavity was found to be empty, and the lining membrane free from disease, but at the upper part, and near to the septum ventriculorum, a group of petechiæ existed beneath the membrane. The right auricle was normal; the left ventricle contained some coagulated blood of a dark hue, which, being removed, showed similar spots on the septum to those seen in the right ventricle. The muscular structure of the outer wall of this cavity was discoloured by blackish streaks and spots.

" On arriving at the farm, I first made an autopsy of another lamb, and found similar morbid appearances to those above described, with the exception of the petechial condition of the heart. My attention was then specially directed to a lamb which was suffering from the disease; the symptoms were analogous to those named by the bailiff. I ordered its removal to a well-ventilated shed, prescribed some aperient medicine, and gave directions for it to be kept apart from the others. On minutely inspecting the flock, I could not discover any indications of ill health in the animals; but concluding that the quality of the food was mainly concerned in producing the attack, I determined on making a complete change both in the management and feeding of the sheep. I therefore had them turned on to a common where the herbage was scanty, and where they could roam at liberty, and ordered that they should be carefully watched. On the succeeding morning another ewe was found dead, which was also forwarded for my inspection. This I sent to the college, when the following lesions were discovered:—

"The abdomen was found to contain a large quantity of fluid of a sero-sanguineous character, and venous congestion of all the abdominal viscera existed to a considerable extent, some parts being nearly black. The vena porta and the contiguous portion of the posterior vena cava were distended with coagulated blood. The spleen was likewise much enlarged. The biliary ducts, gall bladder, and ductus communis choledochus were full of bile; and the liver, as in the lamb, was large and dark in colour, from repletion of its vessels and ducts. A small quantity of fluid was found in each pleural cavity; the lungs were much congested, but no structural disease existed; the pericardial sac contained about its usual quantity of fluid. The heart had an unnaturally large appearance. On laying open its right side, both cavities were found to contain a large quantity of coagulated blood, which likewise extended into the large vessels connected therewith, particularly the anterior and posterior cava and coronary veins. The right auricle, when freed from its contents, also exhibited ecchymosed spots in its muscular structure beneath the lining membrane. The left side of the heart contained but a small quantity of blood, which was likewise of a dark black colour, but showed no marks of structural disease."

Other diseases are occasionally termed red water, and, in the majority, the leading symptom amongst sheep, as amongst cattle, is the redness of the urine.

Professor Murray, of the Cirencester Agricultural College, speaks of red water in ewes as characterized by jaundice, port wine colour of urine, and loss of condition; as the disease advances, the anæmic condition is very marked, the heart's action is frequent, sometimes up to 140 per minute, and its sound loud, the respirations are also rapid, and the animal soon sinks from sheer exhaustion.

The pallor, or yellowness of the tissues of the body, softened condition of the liver, and scanty quantity of blood, are the characteristic post-mortem appearances.

Treatment consists in giving nutritious food and tonics, both mineral and vegetable.

Delafond describes a variety of these blood diseases under the name *diarrhœmia*, and their general characters are— breaking up of the blood, ecchymoses, secretions tinged with blood; and as an illustration of the cause of these conditions is defective nutriment, he alludes to the appearance of such symptoms in 1851 amongst some horses purchased for anatomical purposes, and which were kept several days without food.

BLOOD DISEASE IN LAMBS.—NAVEL ILL.—PYŒMIA AGNORUM.

This, the so-called " new malady in lambs," was first observed by me in 1861. It had destroyed some lambs in various farms for a year or two previously, but it only manifested itself in great severity in the year that I was first consulted, owing to an outbreak in Northumberland. In England, and especially in Hampshire, Wiltshire, and the county of Gloucester, the disease has been very prevalent and very fatal for some time past. In Scotland the disorder has been seen principally in Berwickshire. The seasons during which it has been most rife, have been remarkable for the good condition of the ewes, the heavy crop of lambs, and abundance of food. Deaths have risen rapidly in a flock from one to a score in the first week or two of the lambing season, and instead of diminishing with mild weather and an improving crop of grass, it has killed so rapidly that I have been frequently informed by the farmer, that he counted the number of deaths up to thirty or forty, but they occurred so rapidly then that he " got confused in his calculations."

I have usually found that the flocks affected with this disease have been kept in a confined space in winter. The ewes have been fed well, and not been allowed to move about sufficiently. It was thought at one time that contagion was one of the causes favouring the spread of the disease. This arose from the death of all lambs which were caused to suckle foster mothers. As twins and triplets are common when this disease prevails, many of the young ewes that can scarcely be supported by their own mothers, are placed with the ewes which had lost their offspring. To make the ewe take to the lamb, the skin of the dead one is put on the latter, and this is attended with bad results. The lamb thus covered up often dies, and it was thought that this depended on the young animal catching the prevailing disease. It turned out, however, that the real cause of death was suppressed action of the skin, or poisoning the blood of the otherwise plethoric lamb.

The symptoms of the disease are sudden staggering and drooping look of the finest lambs. They are sometimes costive, and at others purged. The navel is then felt to be swollen and flabby. The eyes get yellow. The animal cannot stand, but if lifted to its dam, attempts to suck. It dwindles, and dies in from a few hours to a week.

After death, dark blood is found in the viscera; the umbilical veins are swollen, the liver is engorged, and studded with multiple abscesses, and there is often yellowness of the tissues of the body generally.

Treatment consists in placing the flock on bare keep, and either giving purgatives or doses of neutral salts to the ewes.

As to prevention, moderation in feeding is the great secret, so as to keep up the condition of stock, without having an excess of internal fat, and an extraordinary richness of blood. All the organs of these animals must be kept in activity, by appropriate exercise. It is bad for all breeding animals to be kept too quiet. They should be made to run about, even after they are drafted out in the order they are to lamb. It is impossible to be too careful with this. Some have suggested giving common salt to the ewes in food. This is a useful agent in rot and other disorders attended with an impoverished state of blood, but it is very injurious in animals that are plethoric. If any medicine is required in consequence of the excessive good condition of a flock of ewes, it should consist in epsom salts to purge them, or nitre to cool them. They will lick up nitre greedily, and it is an admirable preservative.

JOINT DISEASE AMONGST CALVES AND LAMBS.

In conjunction with the blood disease already described, and sometimes independently of any such affections, young domestic ruminants suffer at, or immediately after, birth, with a form of arthritis, dependent partly on a scrofulous cachexia, but having most of the characters of a rheumatic disease. This arthritic disease has been much on the increase of late years, and especially amongst lambs. Mr Robertson, of Kelso,* who has studied this disease with his usual care and intelligence, considers that its causes are constitutional and local. The constitutional and predisposing causes are a scrofulous and a rheumatic taint. Certain exciting causes are essential to the development of the disease, and "animals which have undoubtedly inherited either a scrofulous or rheumatic diathesis may, under favourable conditions, escape being affected; while others, less fortunately circumstanced, may fall victims, or, as in certain cases, both these tendencies are associated in the same animal."

Mr Robertson declares it as his opinion, that calves are most frequently affected by true rheumatism. The animal is seized when some weeks old, and it is observed to be very lame. The pain experienced is evidently intense, and general fever high. The temperature of the body is increased, the visible mucous membranes injected, and the pulse frequent and full. The appetite is capricious or lost,

* See the *Edinburgh Veterinary Review*, vol. v. 1863, p. 529.

the animal is costive, its urine high-coloured, and the affected
joints are hot, tender, and swollen.   The capsules of the
joints are distended, and there is more or less general tumefaction.   The local symptoms may disappear from one joint,
and attack another, or several.   The joints undergo degeneration if the disease continues.

When calves are scrofulous and seized with joint disease,
Mr Robertson finds that the stifle joint is chiefly involved.
The disease is more one of the ends of bones—the epiphyses
—than of the synovial membrane, or other structure of the
joints.   The joint is very tense, hard, and swollen.   There
is a tendency to tubercular deposit in the bone, and softening.

In lambs the form of disease is chiefly of the second
variety above described, the animals are either born with the
diseased joints, or show symptoms shortly after birth.   Mr
Robertson says, that the shepherd is apt to remark that
"this or that lamb cannot live, it is pocking at the navel."
The belly is pendulous from the presence of a turbid fluid in
the peritoneal cavity; in this fluid are floating shreds of unhealthy looking fibrin.   The umbilical cord is always much
enlarged, so much so as to attract attention, whenever the
lamb is dropped; it is soft, flabby, and the vessels filled with
very dark coloured blood.   There does not seem the least
inclination to that early change of these structures, into the
well defined ligamentous cord extending to the liver, characteristic of the perfectly healthy animal; instead of this there
is developed a chain of cysts, containing pus mixed with
tubercular matter, extending from the umbilicus to the liver,
this latter organ exhibiting change of structure, and the
presence of pustular and tubercular matter; the omentum
and mesenteric glands occasionally showing like morbid
conditions, from which, as a sequel, we have the presence in

the abdomen of the already mentioned purulent serous fluid, and externally, evident pendulous abdomen.   In all such cases when the constitutional cachexia is so marked and destructive in its progress, we are only able, by the most careful treatment and nursing, to save a small minority; and I am of opinion it is better not even to attempt this, as any that may recover are never remunerative as breeding, and very rarely as feeding animals."

Treatment.—In any cases of this disease which it is deemed proper to treat, the young animals should have acetate of potash, or the acetate of ammonia, given to them in water repeatedly.   The joints must be kept still, and a starch bandage often benefits much, especially in calves a few weeks old.   When the local swelling assumes a chronic character, iodine preparations are needed, or fly blisters.

With regard to preventive measures, Mr Robertson strongly insists on the selection of sound breeding stock, and goes on to say, that although it does not appear that any one has as yet been able to detect with such nicety, as in the human family, those unmistakeable characteristics of a scrofulous tendency, still there are certain points, which as unmistakeably stamp an animal as objectionable, because of an unhealthy disposition, as others, which give him favour in our eyes as of a superior class, and in the enjoyment of the most vigorous health.   I do not suppose there is any one who, knowing even little of sheep, would select, as fit for breeding, an animal with a thin neck, narrow chest, pot belly, narrow loins, tender eyes, very small bone, and fine wool, which is sparingly distributed about the head, belly, and legs."   Mr Robertson alludes also to moderation in diet, and abundance of exercise, which I have insisted on, as a preventive against the blood disease in lambs.

LAMB DISEASE IN AMERICA.

In the spring of 1862, the lambs in all the counties of New York suffered severely from a peculiar malady, described by Mr Randall, in the "Albany Country Gentleman."* The lambs affected had the appearance of a general want of physical development at the time of their birth. Their bodies were small and lean, or if not, they had a peculiarly flaccid feeling, as if the muscles had not attained their normal consistency. The bones generally lacked the usual size. The back and neck were thin, the legs slender, the head small, and the face oftentimes singularly attenuated. When to these appearances was added the not unusual one of a coating of wool and hair much thinner and shorter than usual, the resemblance to a prematurely born animal was striking.

"Some were brought forth so feeble that they never rose to suck. A portion survived for a few moments or hours; others lingered along from two or three days to a week. They were usually dull, made but languid efforts to feed themselves if their dams were at all shy, and many of them would scarcely follow their dams about the yards or fields. Those that survived required extra care, and very few of them attained ordinary size and plumpness, however plentiful their supply of milk.

"Congenital goitre in some instances accompanied the preceding symptoms. In several flocks, a few of the lambs were born with their heads and necks so drawn down, and occasionally also twisted sideways, by the action of the muscles, that they could only suck with difficulty, and by assuming the most unusual postures. In the worst cases, the lambs starved if they did not receive assistance from the

* See the *Edinburgh Veterinary Review*, vol. v. 1863, page 105.

shepherd until they acquired strength to make the unusual
exertions required of them.   In the same or other flocks,
another set of symptoms appeared.   Strong healthy lambs a
week or two old suddenly lost the use of their legs to a
greater or less degree.   Some hobbled about as if lame in
every foot; others could scarcely walk.   A portion grew no
worse, and after a few weeks recovered.   A small number
became unable to stand even when placed on their feet; but
they continued to look healthy, fed heartily when assisted,
and, so far as my own immediate observation extended, most
of them gradually recovered when the weather became settled
and warm.

" The local visitations of the epizootic of 1862 were quite
capricious.   While many flocks of sheep of all grades in this
(Cortland) county wholly escaped its effects, much the larger
number were losers by it, in proportions varying from 10 to
90 per centum—or practically to 100 per centum, for the few
that recovered in badly diseased flocks were of little value.
The average loss in the *larger* infected flocks was, I think,
about 50 per centum.   My flock lost 40 per centum, my
son's 70, and a neighbour's 90."

Referring to the causes Mr Randall says:—"That our
flocks of sheep in New York were unusually confined during
much of the winter of 1861-2, is certain.   Uncommonly
deep snows fell about the first of February, and though they
wasted away towards spring, their hard crusts prevented
sheep from straying from the immediate vicinity of their
stables.   Many flocks scarcely moved fifty yards from their
stables during the last ten or twelve weeks of their pregnancy.
Their appetites were kept keen by the steady cold.   The free
consumption of food, inaction, and advancing pregnancy, increased their flesh, and these causes reacted and rendered
them perfectly contented in their confinement.   Many flock-

masters have remarked to me that they never before saw their sheep so quiet, so disposed to remain constantly in their stables, and so fleshy towards spring. Having eaten, they lay most of the time in their bedding until they again rose to eat. Flocks accustomed to run in pastures in the winter, and to dig down to the grass, were of course entirely cut off from their usual supply of succulent food."

At the conclusion of his paper Mr Randall refers to the same points I have so often insisted on as to the overfeeding of ewes, and keeping them in an inactive state. He says:— "I believe that I have seen the fact repeatedly established, that it will not do to let pregnant ewes obtain green food by roving about the fields and turnip patches, for the first two or three months of pregnancy, and then confine them rigorously to a small yard and dry food. Some farmers habitually do this, but I never saw it done with impunity in a large flock. In winters unfavourable to sheep, it often leads to a wholesale destruction of even the grown animals.

PARASITIC DISEASES.

The subject of parasitism has within the last twenty years afforded scope for the most interesting researches.

In former times there was a tendency amongst medical observers to attribute many of the most fatal and most common diseases of man to the influence of animal or vegetable parasites, which were often supposed to generate spontaneously under stated circumstances. As Leuckart says, "There was no severe or dangerous disease which parasites, and especially intestinal worms, were not supposed to induce."[*]

* Die Menschlichen Parasiten und die von ihnen Herrührenden Krankheiten von Dr Rudolf Leuckart, Leipzig, 1862.

A reaction occurred amongst pathologists, and as many products supposed to be parasitic proved to be nothing of the
sort, it was supposed that entozoa existed in the bodies
of men and animals for some wise purpose, and excited the
secretions, favoured digestion, &c. Amongst veterinarians
Bracy Clark advocated such views early in the present
century, and went so far as to recommend horsemen to
give their horses some of the germs of œstrus equi, that
their stomachs might not be deprived of the healthy stimulus
which they enjoy in a state of nature from the usual system
of propagation of these parasites. Bracy Clark thus advocated doctrines which had been defended by no less eminent
naturalists before, such as Götze and Abildgaard.

It was supposed by others that parasites developed in animals previously diseased, and that a predisposition had to be
acquired by a certain state of ill health for the production of
any parasitic malady.

We now know that parasites are not generated in certain
morbid conditions, and do not exist in animals to excite the
normal functions of their organs. They are offensive products foreign to the bodies of the men and animals they
afflict, and dependent entirely for their development on the
introduction of germs into bodies suited to their growth, protection, and reproduction.

A few parasites exist in or on all human beings and animals, but certain parasitic animals and vegetables induce
actual disease, and often diseases of a very fatal nature.

The manner in which entozoa injure and destroy is not
always the same. Some induce disease and irreparable structural changes in important organs, from their mere growth
and multiplication in those organs. Thus the brain of the
ox or sheep is destroyed by coenuri and echinococchi. The
latter parasites and flukes lead to destruction of the livers

especially in sheep, and tens of thousands of these animals are annually destroyed by the distoma.

A variety of diseases are induced according to the manner in which the parasites lodge in an organ, or according to the peculiarities of the organ itself. Thus echinococchi not unfrequently induce cardiac tumours in the lower animals, attended by all the symptoms of chronic heart disease, and ending in sudden death. Parasites in the cranial cavity lead to paralysis, wasting of the body, and many complications ending also in death.

Tubular organs are obstructed by parasitic accumulations. Thousands of the calves, sheep, fowls, pheasants, &c., are annually suffocated by round worms in their wind-pipes. Obstructions of the alimentary canal occur in young animals from the accumulation of ascarides.

I have said that parasites induce ill effects from the manner in which they lodge in an organ. The trichinæ afford us an excellent example, in penetrating the sarcolemma, and taking the place of the active muscular elements, as described by Leuckart.

Leuckart has spoken of the usually accepted view, that parasites injure by impoverishing the blood of their victims. He has made an interesting calculation on the subject. He says, a tapeworm (botriocephalus latus) of 7 metres in length, weighs about 27·5 grains. It may, during its growth, lasting as it does from five to six months, require from four to six times its weight in nutritive material, but that is of no importance to a man. Greater losses are sustained by children when large numbers of ascarides accumulate in the intestines. The only instance of a parasite killing by draining the blood of man is the blood-sucking anchylostoma duodenale (strongylus quadridentatus), which attaches itself to the mucous membrane of the intestine of the Egyptians and other orien-

tal people, and in such numbers that on opening the intestine
it appears covered with leeches.

So far as my own inquiries extend as to parasites in the
lower animals, none kill by merely draining the system of
blood. I shall refer elsewhere to this supposed action of
distoma hæpaticum.

Parasites are living and moving bodies, and in their peri-
grinations through the system, or in their movements in a
part in which they are lodged, they induce great derangement, and may kill. I have witnessed this in my experiments with coenuri, and when many germs are introduced
into the system of pigs and calves, &c., for the development
of hydatid disease, deaths are frequent when the embryos
bore through the tissues. In the pig death occurs from the
piercing of the intestine by echynorhyncus gigas, &c.

Leuckart refers particularly to the injurious effects of the
movements of parasites. They induce an irritation which is
followed by congestion and inflammation varying in intensity
according to the number of parasites, and the rapidity of
their movements. He adds that " the most striking example
of the truth of these statements is afforded by the trichinæ,
which, on their passage into the intestinal canal, induce a
malignant enteritis with the production of false membranes,
and lead to appearances which have a great resemblance to
those of typhus. This happens, at all events, when the number of imported parasites is great, amounting, perhaps, to
upwards of 100,000, as is not rarely found after the eating of
trichinous meat. I have seen a corpse in which half-an-
ounce of flesh contained about 300,000 trichinæ. In other
cases the direct results of the parasitism are milder, but
always under the form of a congestive state and catarrhal
affection."

Not unfrequently parasites induce indirectly a derangement of an important organ. We have instances of this in the epileptic seizures or other convulsions of children and of young animals suffering from intestinal parasites.

The parasites, I have to refer to, belong to the three orders of cystocestoid or tapeworms, nematoid or round worms, and tremadote or sucking worms. Of some of these, particularly of tapeworms and sucking worms, it is characteristic that in their development they pass through a non-sexual stage, during which they may infest different animals from those in which they dwell during the sexual and reproductive stage of their existence. And thus the same parasite may kill more than one animal. Human beings derive most of their parasites from the domestic quadrupeds. Leuckart says, " The chief result of our observations on the life history of the helminthoid animals is to the effect that by far the greater number of these creatures live in their various conditions in different animals. Applying this conclusion to the human parasites, we find that in all probability the greatest part of our entozoa are derived from animals. It is the animals with which we come most in contact, viz., our domestic animals, and especially those we eat, that communicate parasites to us— . . . "The justness of this conclusion is demonstrated without doubt by observations and experiments. The domestic animals furnish us, in fact, with the greater number of parasites, but under different circumstances. The parasites which we derive from the animals we eat, such as the tapeworm and the trichina, belong to the developed intestinal worms. We acquire them in their young state, the tapeworm as hydatids, and the trichina as an encysted muscle worm, and both from pigs, which are the animals that mostly give us the eggs and embryos of their entozoa, which then develope in our bodies in their early condition. Of the encysted parasites the dog, above all others, supplies us with germs. It is it that favours the spread of pentastomum denticulatum, cysticercus tenuicollis, and echinococcus, from the development within the nasal sinuses of pentastomum tænioides, and in its intestine of tænia marginata and T. echinococcus. Also the muscle trichinæ of men may in some cases, especially when they are few in number, be communicated from dog to man."

Of the internal parasities, or entozoa, affecting the domestic quadrupeds, there are many, and have been classed under two general heads—*infusoria* and *worms*.

The infusoria are constantly found in the bodies of animals. Sometimes they engender disease, and no more interesting illustration of this can be adduced, than that of the Bacteria, belonging to the family of Vibrios, and which have usually been described as rigid, filiform animalcules, moving in a vacillating, rather than in a serpentine, or undulating manner. These parasites vary from 2 to 5 millimeters in length. Bacteria were supposed to develope, and live principally in putrid liquids, but M. Davaine has found them in the blood of animals affected with splenic apoplexy, and has observed, that putrefaction destroys the infusoria as well as the blood. They can be transferred from one animal to another by inoculation, and multiply with very great rapidity in the vessels of their new host.

The Bacteria kill the animals they invade, and if the blood containing them is rapidly dried, the infusoria are thereby preserved, and can resist the temperature of boiling water. This explains how in some arthracic diseases, the flesh of animals may prove deleterious after being cooked.

Of the family of Monads, it is asserted, on the authority of Leuwenhœck, that Cercomonas urinarius is frequently met with in horses' urine when fresh.

In the stomach and intestine of the herbivorous quadrupeds,

infusoria usually abound, and special attention has been paid by Leuckart to the occurrence of Paramæcium coli in the colon of the pig. Leuckart has found the parasite very constantly and in large quantities in the alimentary canal of this animal, and has been inclined to regard its occurrence in man as the result of accidental transmission from the pig.

Various infusoria have been occasionally seen in putrid discharge from suppurating wounds.

Worms are classified into tape and cystic worms, sucking worms, and round ones, as follows:—

Cystocestoid Worms.—Bladder and Tape Worms.

A. TÆNLÆ.

I. TÆNIA MEDIOCANNELLATA of man.
  Cystic form in muscles of ox.
II. TÆNIA SOLIUM of man.
  Cystic form is the measle or cysticercus cellulosæ of the pig.
III. TÆNIA SERRATA of dog.
  Cystic form is the cysticercus pisiformis of the rabbit.
IV. TÆNIA CŒNURUS of dog.
  Cystic form is the cœnurus cerebralis of cattle and sheep.
V. TÆNIA ECHINOCOCCUS of dog.
  Cystic form is echinococcus hominis s. veterinorum of man and animals.
VI. TÆNIA CUCUMERINA of dog.
  Cystic form is C. cucumerinus of rabbit. (Cobbold).
VII. TÆNIA MARGINATA of dog.
  Cystic form is the cysticercus tenuicollis in the sheep, pig, &c.
VIII. TÆNIA CRASSICOLLIS of the cat.
  Cystic form is cysticercus fasciolaris of the rat and mouse.

Tapeworms whose cystic forms are as yet unknown:—

IX. TÆNIA NANA of man.
X. TÆNIA EXPANSA of the ox, sheep, gazelle, chamois, &c.
XI. TÆNIA DENTICULATA of the ox in France and Germany.
XII. TÆNIA PLICATA of the small intestine and even stomach of the horse.
XIII. TÆNIA MAMILLANA of the large intestine of the horse.

xiv. Tænia perfoliata of the cœcum, and sometimes of the small intestine of the horse.

xv. Tænia elliptica of the cat, declared by Van Baneden to be the same as T. cucumerina of the dog.

xvi. Tænia infundibuliformis from the intestine of the domestic fowl, duck, swan, &c.

xvii. Tænia proglotina of the common fowl.

In the intestines of domestic and other fowls, the following have also been found: Tænia crassula, Tænia malleus, Tænia lanceolata, Tænia setigera, Tænia sinuosa, Tænia fasciata.

B. Botriocephali.

i. Dibothrium decipiens.   Botriocephalus of the cat, found in the small intestine.

ii. Dibothrium serbatum.   Botriocephalus of the dog, found in the small intestine both of the dog and fox.

Nematoid, or Round Worms.

A. Larval Forms.

One form has been found in the human trachea, and another in a cyst in the kidney of a dog.   Little is known of these nematoid worms.

B. Perfect Forms.

i. Genus Oxyuris.

> Species a. Oxyuris curvala of the horse and ass; found very frequently in the cœcum and colon.

> b. Oxyuris vermicularis of man; found in the large intestine and rectum.

ii. Genus Ascaris.

> Species a. Ascaris lumbricoides of man, and probably of the ox; found in the small intestine.

> b. Ascaris megalocephala of the horse, ass, mule, &c.; found in the small intestine, and sometimes in the stomach and in the large intestine.

> c. Ascaris mystax s. alata of the cat, lynx, tiger, &c., and also of the human subject. (Bellingham, Cobhold.)

> d. Ascaris ovis. This round worm of the sheep has been found only once in Vienna.

> e. Ascaris marginata of the dog; found in the small intestine.

> f. Ascaris suilla of the pig.

Many other species of ascarides have been described, and especially Ascaris vesicularis of the common fowl and turkey; Ascaris dispar of the goose ; Ascaris inflexus of the domestic fowl ; Ascaris maculosa of the pigeon, &c.

III. GENUS SPIROPTERA.

Species a. Spiroptera megastoma of the horse; found in tumors developed at the cardiac end, and in the walls of the stomach.

b. Spiroptera sanguinolenta of the dog and wolf; also found in tumors of the œsophagus and stomach. In some countries these ₊parasites are common, and as some observers have found them in rabid dogs, there was a belief at one time that rabies depended on this worm.

c. Spiroptera strongylina of the pig and wild boar; found in the stomach.

d. Spiroptera hamulosa of the common fowl.

e. Spiroptera uncinata of the tubercles of the œsophagus.

IV. GENUS TRICHINA.

Species a. Trichina spiralis ; in the muscular fibres of man, pig, ox, rabbit, and other domestic quadrupeds.

V. GENUS TRICHOSOMA.

Species a. Trichosoma plica of the dog; found in the urinary bladder.

b. Trichosoma brevicollis of the goose.

c. Trichosoma longicollis of the domestic fowl.

VI. GENUS TRICOCEPHALUS.

Species a. Tricocephalus dispar of man.

b. Tricocephalus affinis of the dog ; found in the cœcum of ruminants.

c. Tricocephalus depressiusculus of the dog; found in the cœcum.

d. Tricocephalus crenatus of the pig and wild boar; found in the large intestine.

VII. GENUS FILARIA.

Species a. Filaria lacrymalis of the horse and ox; found in the rabbit.

b. Filaria papillosa of the horse, ox, and ass; found in the globe of the eye, said to be in the anterior chamber, but usually in a cyst within the cornea.

c. Filaria immitis of the dog; found in the heart.

d. Filaria trispinulosa of the dog; found by Gescheidt in the capsule of the crystalline lens.

VIII. GENUS DOCHMIUS.

    Species a. Dochmius hypostomus of the sheep, goat, and other
        ruminants; found in the intestine.
       b. Dochmius tubæformis of the cat; found in the duodenum.
       c. Dochmius trigonocephalus of the dog; found in the
        stomach and intestine. A variety declared to exist
        in the right side of the heart.

IX. GENUS SCLEROSTOMA.

    Species a. Sclerostomum armatum of the horse. Found in the
        intestine, in the arteries, and sometimes in other
        cavities of the body.
       b. Sclerostomum tetrachantum of the horse. Found in the
        cœcum and colon.
       c. Sclerostomum dentatum of the pig. Found in the
        cœcum and colon.
       d. Sclerostomum syngamus from the trachea and bronchial
        tubes of fowls.

X. GENUS STRONGYLUS.

    Species a. Strongylus radiatus of the ox, and several other rumin-
        ants. From the small intestine and colon.
       b. Strongylus venulosus of the goat. Found in the small
        intestine.
       c. Strongylus micrurus of the horse, ox, ass, &c., in-
        festing the air passages.
       d. Strongylus filaria of the sheep, goat, camel, &c., also
        infesting the air passages.
       e. Strongylus paradoxus of the pig; found in the trachea
        and bronchial tubes.
       f. Strongylus filicollis of the sheep; met with in the small
        intestine.

XI. GENUS ANCHYLOSTOMUM.

    Only one species of this genus — anchylostomum duodenalis, has
    been described, and it occurs very rarely, if ever, in the lower
    animals. It usually infests the duodenum of man.

XII. GENUS EUSTRONGYLUS.

    Species. Eustrongylus gigas of the horse, ox, dog, &c. Found in
        the kidneys, bladder, and areolar tissue, beneath the
        peritoneum.

Trematode, or Sucking Worms.

I. GENUS MONOSTOMA.

Species. Monostomum leporis, attacking the rabbit, and found in the peritoneal cavity.

II. GENUS DISTOMA.

Species a. Distomum hæpaticum. This parasite attacks ruminants principally, but it has also been found in the horse, ass, pig, rabbit, and man, in addition to many wild animals.

b. Distoma lanceolatum of domestic ruminants, besides the pig, cat, and rabbit. Found in the biliary ducts.

III. GENUS HOLOSTOMUM.

Species. Holostumum alatum of the dog. Found in the intestine.

IV. GENUS AMPHISTOMUM.

Species a. Amphistomum conicum of the ox and sheep. Met with in the stomachs.

b. Amphistomum crumeniferum of the ox.

c. Amphistomum explanatum of the ox. Found in the liver.

d. Amphistomum truncatum of the cat.

Aranthocephalis, or Armed Worms.

GENUS ECHINOCHYNOUS.

Species. Echinochyncus gigas of the pig; found in the intestine.

Aconthotheci.

GENUS PENTASTOMUM.

Species a. Pentastomum constrictum of ruminants and man.

b. Pentastomum tænioides of the dog, horse, mule, sheep, &c. Found in the frontal and ethmoidal sinuses.

MEASLES IN THE PIG.—SCALESIASIS; CACHEXIA HYDATI-GENA; LADRERIE, FR.; FINNEN-KRANKHEIT, GERM.— TÆNIA SOLIUM IN MAN.

This disease of swine has been entirely overlooked by veterinarians in this country, and it has been only since the researches of V. Siebold and Küchenmeister that British physicians have ascertained the frequent existence of parasites in pigs, which, on reaching the human intestine, develope into tæniæ.

The very inappropriate term "measles" is applied to that morbid state induced by the presence of cysticercus cellulosæ in the muscular structures of swine. It is a purely parasitic disease, and depends for its origin on the introduction into the system of the pig of the mature and fecundated ova of tænia solium.

The process of development has been carefully watched by many observers. The embryos of the tapeworm are globular and armed with spines, which pierce, by working in a horizontal plane from within outwards, the mucous membrane of the alimentary canal of the pig. They penetrate the tissues, and are washed through the larger vessels by the blood current until they reach their destination in the muscular structures. A very large number of the embryos are thus dispersed, but only in young animals. They cannot find their way through the tissues of adult pigs, and any experiments performed with animals above a year old fail as a rule. This, as we shall afterwards see, is the same with other parasitic diseases.

Pigs are said to be born measly, and one of the most constant means whereby the disease is propagated is by breeding from measly sows. French veterinarians long since noticed that if a measly sow was bred from, all her produce was measly, and similar observations have been made in this country by the bacon factors.

If pigs are born healthy they cannot have fully developed cysticerci in their flesh under two months and a half. From 30 to 40 days after the introduction of the germs into the body of the pig the parasites vary in size from one to four millimetres. They consist in small cysts, or bladders, containing a clear fluid, and in the wall of the cyst there are many distinct vessels. A rudimentary head soon appears, and then a row of hooks, and, lastly, suckers around them develope. Each cysticercus is enveloped in a cyst, its body grows, and is in reality drawn into the bladder, with which it is continuous at the opening, so that the vesicle proper to the animal is in reality the tail. Cysticerci continue to grow for four or five months, but then remain stationary, and, although occasionally killing the animal in whose flesh they have accumulated in countless numbers, they usually have no means of escape until the natural term of the pig's existence is at an end, and then they pass into the bodies of human beings.

I have seen pigs, in whose flesh cysticerci abounded, in apparently the most perfect health, and very fat. Indeed, it is necessary to examine an animal closely during life in order to determine if it be measly. The parasites are usually situated superficially under the tongue, and may be felt on the inner side of the eyelids. In very severe cases the neck is swollen, there is difficult breathing, and a hoarse voice. It is a mistake to suppose that measly pigs have red spots on the skin, or any sign of cutaneous eruption.

After death the presence of cysticerci is easily seen in the different muscles, and in the internal organs. It is especially when a pig is cut into two halves, and the muscles of the neck are cut through, that the greatest mass of these parasites is exposed. The pork butchers usually make an incision into the psoas muscles to determine if a pig is measly. The presence of many cysticerci in the flesh leads to an open condition of the texture favourable to the imbibition of fluids, and for this reason measly pigs are easily pickled.

My inquiries indicate that measles prevails to a much larger extent in Ireland than in Britain, and may be regarded, in fact, as enzootic in the former country. I have been informed by a Wiltshire bacon factor that not one pig in a thousand reared in England or Scotland is found diseased, whereas the Irish pigs suffer much from this disease, and some years to an extent of six, seven, and eight per cent. The Irish have an adage that every pig has its measle, and if I consider what number of animals have a few cysticerci in their flesh,the per-centage of measly animals is far higher than above stated.   When we speak of a measly pig there is an accumulation of many hundred such parasites in the animal's body.

I found that the malady was most prevalent in those counties in Ireland where pigs are reared in small lots by poor people.   The disease has diminished considerably of late years, in consequence of the pigs being fed in larger numbers by farmers.   I found that measles were very rife in some parts of Cork, in Limerick, Tipperary, and Queen's County.   Of the counties in the province of Ulster, Monaghan is by far the greatest sufferer by this disease, and I regret that I have not had an opportunity to follow out my inquiries further as to the causes which lead to the extraordinary losses by this disease in special counties.

It is certain, however, that those pigs suffer most from measles that live in common with human beings ; that are allowed to roam about at will; and to eat human excrement around the cottages, in the roadside, &c.   A very few people affected with tapeworm discharge joints enough to contaminate an immense number of pigs.   Each tapeworm has an average lifetime of two years.   It produces in that time 1,600 joints, and each of these contain 53,000 eggs, making in all 85 millions.*   Every egg is capable of developing into a cysticercus, but fortunately the great majority of the joints of a tapeworm are destroyed.   Were they not, every pig would soon be measly, and every man, woman, and child suffer from tænia solium.

* Leuckart, *loc. cit.*, p. 83.

MEASLES IN CATTLE.—TÆNIA MEDIOCANELLATA (KüCHEN-MEISTER).

Recent researches by Dr Leuckart demonstrate incontestably that there is a form of tapeworm, not unfrequently confounded with tænia solium, which does not originate in man from eating measly pig, but from eating imperfectly cooked veal and beef. In many parts of the world a hydatid prevails amongst cattle, which developes into tænia mediocanellata in the human intestine. That hydatid is found in many parts of Europe, and probably exists occasionally in this country. Dr Cobbold has a specimen of tænia mediocanellata in his collection, obtained from Sheffield, and he informs me that we shall probably find that this variety of tapeworm is not at all rare in this country. There is a specimen in the New Veterinary College Museum, for which I am indebted to Dr Keith of Aberdeen. Leuckart quotes an observation which interests us as Englishmen. He says that Knox observed a tapeworm epidemic during the Kaffir war in 1819 amongst the English soldiers, due to their being fed on unsound beef. Abyssinians are affected with this disease, and observations have been made in Germany and Russia as to the occurrence of tænia mediocanellata amongst children, fed—"aus diätetischen Gründen"—on raw beef.

Dr Leuckart has succeeded in inducing measles in the calf, by feeding it with joints of tænia mediocanellata.

As hydatids prevail to a very extraordinary extent amongst cattle and sheep in this country, it is very important that a carefully conducted inquiry should be prosecuted, with a view to determine the existence or non-existence amongst us of tænia mediocanellata, and the cysticerci which induce them.

Hydatids of the Liver in Animals, Cysticercus Tenuicollis.—Tænia Marginata in the Dog and Wolf.

The pigs in Ireland, and both cattle and sheep throughout the United Kingdom, suffer to a very great extent from hydatids in their livers. Amongst these cystic parasites we find a large number of the species Cysticercus tenuicollis. These cysticerci are apt to take up their abode also in the internal organs of man, and it is probable that they often lead to the development of cysts supposed to have been due to the presence of echinococchi, and I am inclined to attribute to this parasite the cystic tumours which Dr Brinton, and even Dr Gairdner, consider arise in human beings eating raw or underdone animal food. Human beings suffer from these cysticerci under circumstances similar to those which lead to the development of cysticercus cellulosæ, and which I have before alluded to.

There is no doubt that eggs of the tapeworm developed from cysticercus tenuicollis in the intestines of the dog, will develope into hydatids in the mesentery and liver of human beings, as it does, according to the experiments of Luschka, Leuckart, and others, in the organs of the domestic quadrupeds.

It is of the greatest importance that careful and extended inquiries should be made as to the prevalence of these cysticerci in animals. It is evident, from the observations of Küchenmeister and others, that many individuals of these species, forming extensive cystic tumours, are to be found in pigs, and not unfrequently there has been a confusion between cysticerci and echinococchi. Thus, in Ireland, the endemic cystic disease appears to be due to both these hydatids.

ECHINOCOCCUS VETERINORUM, TÆNIA ECHINOCOCCUS.

Numerous cases have come under my notice of disease in horned cattle, sheep, and pigs, induced by this parasite. Von Siebold has shown that tænia echinococcus lives in the dog's intestines, and thousands of thread tapeworms may exist in an animal.

It has been a much agitated question whether there are several species of echinococchi. The earlier observers believed in two species, ecchinococcus hominus, and ecchinococcus veterinorum. Weinland says: "As to the difference of the two species there can be no longer any doubt, since the investigations of Küchenmeister and Leuckart." The latter author, however, says, at page 330 of his new work, referring to other authorities on the subject: "They thought themselves so much the more justified to make this difference, inasmuch as the first (E. hominis) are characterised by the presence of secondary and tertiary bladders (tochter und enkelblasen) within them, whereas the others usually present a simple cystic form. But it is known that the multiple form of Echin. hominis occurs also in the domestic animals such as the horse, pig, &c., and *vice versa*, it is not rare to find in human beings the simple forms of E. veterinorum. Sometimes, moreover, both forms of echinococcus are found in the same individual." Leuckart refers to the simple form of echinococcus hominis in some cases, and to the complicated form, E. veterinorum, in others; and, after reference to the shape and number of hooks, concludes by saying: "Naturally under such circumstances I can no longer participate in Küchenmeister's views, that there are two species of echinococcus, and the forms of this parasite indigenous with us are only varieties of a single species, whose fully developed condition is to be met with in our dogs."* ·

Tænia echinococcus, first seen by Von Siebold in experiments on dogs, is a small tapeworm with only three or four joints, the last of which, in a mature condition, exceeds in size the remaining part of the body. As a rule, their number in the dog's intestine varies from a few to 30 or 40.

I have seen masses of echinococchi, weighing many pounds, appended to the apex of the heart, others connected with the lungs, liver, spleen, kidney, and the last specimens I obtained were in the cranial bones of a bullock. Echinococchi are far more frequent in Italy, where I have seen them in enormous numbers, than in the United Kingdom, but they are very common in this country also.

CŒNURUS CEREBRALIS IN CATTLE AND SHEEP; GID, STURDY, TURNSICK.

The very common disease, sturdy or gid of the sheep, Dreh-Krankheit of the Germans, prevails to an extraordinary extent in all parts of the United Kingdom where sheep are kept. There are districts comparatively free from the disease, and others where there is an annual loss of one and two per score among year-old sheep.

From the very satisfactory explanation of the origin of

---

* I have had numerous opportunities of examining echinococchi from man and animals in Italy, as well as in this country, and have very frequently studied them carefully. I have always referred to my own observations in the lecture room as leading me to differ from those who considered that there were two species of echinococcus; and during the past session, before I had the pleasure of reading Dr Leuckart's admirable work, I entered at length in the class-room on the supposed but imaginary differences between the echinococchi of man and those of our domestic quadrupeds.

this disease, which is afforded us by a knowledge of the source whence sheep or cattle derive coenuri, I attempted to convince the farmers, several years back, as to the real cause of the disorder and the ready means of prevention. As the German zoologists had done, I gave dogs the hydatids from the brains of sheep affected with sturdy, and obtained large numbers of tæniæ. The joints of tænia coenurus thus obtained were given to lambs, and sturdy was induced in them.

In 1859 I drew up tables showing the results of many experiments performed in different countries on this subject, and 41 experiments as to the development of tænia coenurus showed that of about 50 dogs fed on whole or portions of coenuri from the brains of sheep, 33 became affected with tapeworm. As many as 400 tæniæ have developed from one cyst, and the fourth part of one hydatid swallowed by a dog led to the development of 191 tapeworms. In less than a fortnight the tapeworms are observed in the intestines, from a line to two in length, showing no trace of joints or transverse folds; they may attain an inch the third week, and 4 inches the fourth. Worms developed in 155 days are mentioned as being from 2 to 2½ feet in length, but by one experiment it was found that this length could be attained by the tapeworm in less than three months. The tæniæ remain in the small, and obtain exit from the body through the large intestine. They are never expelled whole, but separate proglottides or joints, each of which is charged with many hundred eggs, are evacuated with the fæces.

Failure in the experiment depends on diarrhœa causing the expulsion of the cyst before the heads can attach themselves and grow. Occasionally a disease such as distemper may prevent the retention of the parasites.

To demonstrate that the cerebral hydatids are produced by introducing ova from the dog's tapeworms, 39 sheep and 2

calves received proglottides of tænia coenurus, and out of
these 22 became affected with sturdy. Symptoms of the
disease became manifest from 7 days to 2 and even nearly
4 months after the proglottides had been swallowed by the
sheep. The rapidity with which sturdy developes is almost in
direct ratio with the length of time, within certain limits,
that the joints of the tapeworm have been exposed to the air
and moisture. The tardy manifestations of symptoms in some
cases probably depends on the ready adaptation of the brain
to the developing cysts. The number of coenuri found in
the brain varied from 4 to upwards of 200. They were generally distributed throughout the substance of the brain.
Encysted and undeveloped embryos are found frequently in
the muscular tissue, especially of the œsophagus, intestine,
diaphragm, and heart. The experiments fail if proper attention be not paid in procuring mature joints of tænia
coenurus.*

It is a fact that sturdy rarely affects sheep above two years
of age, usually lambs under a year old; it is more frequently
seen in some breeds, such as among the Cheviots, than in
others, and affects enfeebled animals, more especially in the

---

* Until 1853, the period of Küchenmeister's experiments on the
transmission of cœnurus cerebralis, many were the supposed causes of
sturdy. As a matter of curiosity, a few may be referred to, and I shall
mention those which have been most believed in by farmers and shepherds :—Lullin and Gerike thought sturdy was serous apoplexy, or
dropsy of the brain, from violent blows. Many have believed that
humidity produced the disease, and Navières suggested that a fly
deposited eggs in the brain by perforating the skull, and the eggs
developed into the hyatid met with in the sturdy. The Ettrick Shepherd stated that sturdy was due to cold affecting the sheep's loins, especially during windy and rainy winter seasons. We have been asked how
to explain the prevention of sturdy by covering the sheep's loins. Admitting that occasionally this may protect them, we shall afterwards

autumn and winter months. I find, however, that in some districts there is greater prevalence of sturdy in summer. This occurs when, during the hot months, sheep are kept on unenclosed pastures on hills where they must be constantly "herded," whereas during the winter the flock is transferred to enclosed fields, and dogs are more or less removed from them. Sturdy will always be found to prevail on farms with open pastures, where flocks constantly need the guardianship of shepherds *and dogs,* or on enclosed farms where sheep are fed on turnips, confined daily within limited space, *with one or more dogs amongst them.* These are the conditions favourable to the development of sturdy, and they are those favourable to the dissemination of tapeworm eggs by dogs, and the penetration of the eggs in the bodies of the sheep. These eggs find a favourable nidus in the cerebral mass of the lamb, and they there develope into the coenuris cerebralis.

Sturdy is occasionally confounded with other diseases ; and my attention has sometimes been called by farmers of great experience to a sheep presenting certain anomalous symptoms, which, though distinctly due to the presence of the

show that all conditions calculated to favour the healthy and robust state of the sheep will prevent the introduction and development of parasites in the body, not excepting the cœnuris cerebralis. Fromage de Feugré declares that when lambs are too fat they are most liable to sturdy ; and Reynal only recently advocates the theory of Huzard, that those lambs become affected with sturdy which are born of ewes that have suffered during pregnancy, or that are naturally weak; and, lastly, that the produce of rams of an enfeebled constitution is very subject to the disease. Many shepherds have observed a connexion between the development of sturdy and the presence of dogs amongst the flocks. Many intelligent farmers have a great dislike to dogs amongst sheep, in the belief that by being worried, the sheep become affected with sturdy.

cœnurus cerebralis in the brain, have not been considered those of sturdy. The variety of ways in which the sturdy manifests itself, depends entirely on the number of, and the position held by the parasites in the brain. Usually but one hydatid is found within the skull, sometimes several, and then the symptoms are complicated.

The usual form of sturdy depends on the presence of a hydatid in one of the hemispheres of the cerebrum, or brain proper. The sheep then turns right or left, according to the hemisphere affected. If the bladder be situated between the hemispheres, the head 'is protruded and elevated, and the animal moves in a straight line forwards. Lastly, if the bladder be lodged in the lesser brain or cerebellum, there is defective co-ordination of movement; the creature loses control over the voluntary muscles, there is a peculiar uncertainty of gait; the limbs do not obey the will.

In addition to the above symptoms, there are others which have not been studied as much as they might have been, though of great interest to the physiologist. Such signs are peculiar to different stages of sturdy. We observe that when first affected, the symptoms are very severe; there is much cerebral disturbance from the congestion produced by the presence of the hydatid. As the brain substance yields to the latter and is absorbed—in other words, as the contents of the skull adapt themselves to the parasite,—the symptoms may subside more or less, and a sheep decidedly giddy, stupid, and dull at first, may appear partially to recover; but the growth of the parasite, or any cause favouring cerebral congestion, induces a marked exacerbation of symptoms. But, as Dr Davaine has correctly stated, the vertigo cannot be explained as depending on simple morbid irritation, or looked upon as a symptom of paralysis or incomplete hemiplegia. The attacks of giddiness, the running round and round, become more frequent and are more prolonged as the hydatid grows; the rapidity of movement increases until paralysis is induced, and the animal cannot stand. Many tumours and hydatids of a different species to the cœnurus cerebralis are met with in the brain, but the peculiar symptoms of sturdy are not induced by them.

The cœnurus consists in a bladder provided with a variable number of exsertile heads, and Dr Davaine believes the nervous substance may be excited by the heads, which protrude from the bladder and penetrate the brain substance nearly two lines in depth. Sturdy is, therefore, a phenomenon of excitation of one of the cerebral hemispheres, and Dr Davaine asks if very manifest phenomena of excitation would not result by plunging into the substance of the brain one or two hundred pin-points at a depth varying from one to two lines. As the cœnurus increases in age, the number of heads augments, and the points of contact with the encephalon multiply, and in this way Davaine explains the increase in frequency and duration of the vertiginous attacks as the malady advances.

It is certainly remarkable, that though the echinococcus veterinorum may lodge in the brain of sheep or oxen, it does not produce the characteristic symptoms of sturdy caused by the cœnurus cerebralis, and the probable explanation of this is, that the heads of the former are not exsertile, whereas those of the cœnurus protrude from the distended cyst.

The vertigo observed in true sturdy is altogether peculiar; that is to say, the lamb turns round and round, describing concentric circles, and Davaine states that it has been entirely by false analogy that some authors have admitted the existence of sturdy in man.

Admitting that the cœnurus cerebralis exerts a peculiar influence on the brain, it must be remembered that the "running round" is not a constant symptom. In the early stages, it is often absent, the sensorium and voluntary muscles being more or less affected with dulness and partial paralysis, stiffness of back and awkward gait; there is a peculiar appearance of the eyes dependant on the dilated pupils, the bluish colour of the conjunctiva, and apparent prominence of the eye-ball. Total blindness may result, and the animal feeds but little, cannot follow the flock, strikes against trees, walls, or other obstacles, which it may meet with in moving about.

When the cœnurus cerebralis exists in the cerebellum, a remarkable combination of symptoms may present themselves. The animal advances with its head elevated, can scarcely lift its fore legs, and there is a hesitating movement of all the extremities. Having accomplished the first steps, the creature rapidly advances, occasionally by a succession of imperfect leaps and falls; it then struggles to rise, and may not succeed, or it rolls on its side several times in succession. Emaciation advances, and death ensues sooner or later, but as a matter of certainty, unless the animal is relieved naturally or artificially. The natural method of relief, which is by absorption of the bones of the skull, and evacuation of the hydatid, is very rare, though occasionally a farmer is astonished to learn that a sheep affected with sturdy has struck against a sharp stone, broken its head, and recovered. The explanation of this is, that the skull having become thin, the blow produces a penetrating wound, through which the cœnurus cerebralis may escape. A plan is successfully resorted to occasionally for the removal of the hydatid and cure of sturdy.

Sturdy is occasionally mistaken for functional disorder of the brain, due to impaction of the third stomach, which is a disease of the spring season, of an acute nature, characterized by constipation, delirium, convulsions, and early death, unless the animal be relieved by a brisk purgative.

Sturdy is also confounded with the attacks of the sheep-bot, which is lodged in the frontal sinuses, and produces great irritation, swelling of the pituitary membrane, and discharge from the nose. The animal loses appetite, becomes dull, prostrate, is attacked with convulsions, and sometimes dies.

The Scotch shepherds have become expert in the treatment of sturdy. They feel for the softened part of the skull, pierce the brain with an instrument called a borer, draw off the liquid from within the cyst of the parasite through a canula by means of a syringe, and, if possible, they seize the bladder and draw it out. Many cases are successful if operated on sufficiently early, and when there is but one bladder in the brain.

ROT IN SHEEP: CACHEXIA AQUOSA; THE FLUKE DISEASE; ATTACKS OF DISTOMA HÆPATICUM.

This most destructive disease has attracted more than ordinary attention of late years, owing to its extraordinary prevalence in 1860, and also in the year now closing. Professor Simonds, in a recent essay on this malady,* says, after referring to a number of extraordinary outbreaks:—"From 1830 to the present time several visitations, which were more or less severe, took place. One of these occurred in 1853–54, when many thousands of sheep were swept away, and not only in undrained districts, but also in others of a more healthy character. Since 1830, however, no outbreak can at all be compared to the one of the autumn and winter of 1860. Speaking in general terms, it may be affirmed

* *The Rot in Sheep, its Nature, Cause, Treatment, and Prevention,* by James Beart Simonds. London, 1862.

that all the western and southern counties of England, to-gether with several of the eastern and midland, suffered to a ruinous extent. As in former years, so in this, the attacks of the disease were due to an excess and long continuance of wet weather. Eighteen hundred and sixty will be long remembered by agriculturists, not only as producing the rot among sheep, but likewise for its baneful effects on the root crops, as also on the hay and corn harvests.

"We are acquainted with several instances, in our own immediate neighbourhood on the verge of London, where the losses of sheep amounted from 600 to 700 in a flock. These sheep were principally Welsh ewes, which had been bought at the latter part of the summer for breeding by being crossed with Leicester tups. Some persons lost nearly all, and one in particular, who buys about 800 of these ewes annually, had not more than 40 or 50 which escaped. Tups, wethers, lamb-hogs, and half-breds, alike succumbed to the inroads of the affection. A similar fatality attended the progress of the disease in all other districts. In many parishes in Devonshire where we investigated the malady, and of which Bridgerule may be taken as an example, five-sixths of the sheep perished, or were sold for a few shillings each for slaughtering, to the detriment of the health of the poorer classes.* In the instance thus particularized the losses occurred among the stock of small occupiers, the ill consequences of which were greatly added to by their young cattle being found to be affected by flukes to such an extent as seriously to injure their health later on in the year.

* The Rev. S. N. Kingdon, the resident minister at Bridgerule, reported to the author, that on October 1st, 1860, 492 sheep were existing in the parish as the joint property of several small farmers; and that, by the end of the month, 410 of them had either died or been sold at a price very little above the value of their skins.

" In Sussex and in several parts of Surrey, the fatality was equally great. In the neighbourhood of Eastbourne a flock of about 600 Southdown ewes of great value was completely destroyed. Numerous cases of this kind might be narrated, but enough has been said to show not only the extent of the disease, but that sheep of every description, and placed under different systems of management, equally succumbed. It is much to be regretted that means do not exist whereby the total loss could be ascertained. People are left in doubt as to the amount of food of which they were deprived in one year by this disease alone, and of the efforts which must be made to replace the losses. The time, we predict, cannot be far distant when agriculturists will be convinced, not only of the propriety, but of the positive necessity of making returns, at least of the *losses*, they sustain among their cattle, instead of simply deploring these among themselves. Elsewhere we have drawn attention to this important subject, upon which very much might now be said, if it were not somewhat unsuited to an essay of this kind."

Mr Spooner, in his work on sheep, says, " Though a million of sheep or lambs have frequently been destroyed annually by this disease, in the winter of 1830–31, this number, it is supposed, was more than doubled ; some farmers lost their whole flocks, others a moiety, and many were ruined in consequence. These facts were proved before a Committee of the House of Lords in 1833, and it was there stated by one farmer that he lost £3000 worth of sheep on his farm in Kent, in the course of three months. Even at this time there were 5000 less sheep taken to Smithfield every market-day in consequence of the mortality two years previously, so extensive and general had it been."

My inquiries in 1862 indicate that the mortality in many parts exceeded that of 1860. It has far surpassed it in Ire-

land; and amongst the most extensive sheep dealers in the
midland counties and the south of England I have learned
that the destruction over extensive districts has been almost
unparalleled in their experience.

When I was last in Dublin (Dec. 13, 1862) my advice
was asked concerning this disease, which seems to have prevailed on lands usually quite free from rot, and I learned that
the malady was very destructive in Kilkenny amongst cattle.
Serious complaints have been heard from Clare, Limerick,
Roscommon, King's County, Wexford, parts of Kildare,
Longford, Leitrim, and Armagh. I am quite certain that not
less than 500,000 sheep have this year suffered from rot in
the United Kingdom, reducing them in value two-thirds and
more, and leading to a loss of several hundred thousand
pounds to the country at large.

Rot is a disease of low lands, marshy ground, and wet
seasons. Flooding pastures suffices to render them unsound
for sheep for a season, and this is owing to the dissemination
of distomata in their partially developed condition, and fit for
their term of existence, in the ruminant's liver. Apart, however, from the prevalence of flukes on low land and especially
marshy pastures, we find that sheep do not keep up in condition on soft watery grass. Solid dry food suits the constitution of the sheep best, and during wet seasons we find
rot prevailing to an alarming extent on sound lands, and on
opening the bodies of the sheep very few flukes are found in
their livers. Notwithstanding the existence of flukes in the
liver it is possible to counteract the state of weakness, and
stop the progressive emaciation by rich food, tonics, and
common salt, which do not tend to expel the parasites so
much as to counteract the condition of the system induced
by quality of food the animal has been on, coupled with the
morbid changes in the liver from the presence of the flukes.

Rot developes most readily from the month of June to the month of October.

The fluke, distoma hepaticum, is found in the livers of sheep in a perfect condition, with organs of generation developing or developed, and masses of ova surround the parasites. They are often packed together in scores in saccular dilatations of the gall ducts, and I have seen the most extraordinary specimens of livers, with varicose gall ducts encrusted with cholesterine and other solid principles of bile. The ova, which abound in the gall ducts, pass out through the intestine of the sheep, and fall into stagnant pools, ditches, &c., or are washed from the land during rains into streams. Most of the ova are fortunately destroyed as a rule, but many are hatched, and embryos develope. Steenstrup's investigations on this subject were very remarkable. The embryos were found to acquire great activity, and would move freely, owing to the vibratory cilia formed in their surface. They are eaten by mollusks, the common physæ or limneæ of pools and ponds. The embryos here acquire a sort of hydatid form, are provided with alimentary canal and organs of locomotion. By a process of interior budding *cercariæ* form, which are the young sucking worms, endowed with great activity, and, thanks to a rudder tail, which renders them not unlike a tadpole, they can swim and find their way into water; where they live free until some favourable *crustacean* or mollusk appears, into which they pass by means of spines developed on their head. They lose their tail, and become encysted; their internal organs continue to develope, and on the animal they are infesting being accidentally swallowed by a sheep or other creature, they escape free to pass into the liver, acquire generative organs, and lay eggs for another generation. The metamorphoses here noticed are probably similar for all trematode worms, and are presumed to be those of distoma hæpaticum, whose cercaria form has not been discovered.

Sheep are very liable to suffer from parasites, and in conjunction with the flukes in the liver, we usually find parasites in the lungs and parasites in the stomach. Mr Simonds refers to having recently "brought to light another and a fruitful cause of the death of sheep of all ages," with symptoms "remarkably akin to those of rot," and due to "the existence of an undescribed variety of worm of the class *filaria* within the abomasum,—the digestive stomach." The truth is, that the parasite Professor Simonds refers to, from the brief notice he gives of it, is one which has been frequently referred to before, and is noticed in all German, French, and Italian veterinary works which are at all up to date in matters of science. Bellingham long since noticed the occurrence of strongylus contortus in the sheep in Ireland. He found it in the small intestine, but it is as a rule found in the fourth stomach. It was first described by O. Fabricius in Denmark, who stated that the head of the worm was armed with cilia, probably the barbs which Mr Simonds has noticed. The German authors refer to the disease induced by the gastric parasites in sheep as a "Magenwurmkrankheit." Spinola calls it Magenwürmerseuche, or strongylogenesis ventriculi, and characterizes the disease as "eine cachectische herdekrankheit."

It is, moreover, in the condition of system noticed in sheep rot that many other parasites prey on the bodies of living animals, and echinococchi, cysticerci, &c., are not uncommon in rotten sheep.

Symptoms of Rot.—A flock placed on damp land, or a flock purchased from a country where it has contracted rot, appears to thrive well, lays on fat, and promises to turn into good mutton. Inactivity and dulness are soon apparent. In some cases the disease is rapid in its course, and this season (1862) a large number of sheep have been killed very quickly on lands usually reputed as very sound. Pallor of the visible mucous membranes, wasting, &c., could be seen in these sheep, but only to a moderate extent, and they have died very suddenly. After death the liver has been found greatly enlarged, its peritoneal surface often adherent to the diaphragm and other abdominal organs, and few flukes contained in the liver. The small quantity and pale character of the blood indicate, however, the real condition of the sheep.

As a rule rot progresses at first in an insidious form; the flanks get hollow, the back rigid, and there is a decided yellow colour of the eye, and, where visible, often of the skin ; the fleece drops off in patches; the belly enlarges ; the back droops; and there is a disposition to dropsical swellings in different parts of the body. There is frequently an insatiable thirst as in other dropsical diseases, the pulse is frequent and very feeble, the heart-beats active, and anæmic murmurs are heard ; the breathing becomes quick and short, there is a slight cough, most marked in all cases complicated by the presence of strongyli in the air passages.

The most remarkable of the dropsical swellings is around the throat. A sheep thus affected is said to be *chockered.* The alimentary canal is disturbed, and, with the quantity of liquids drunk, diarrhœa is apt to supervene. Weakness and listlessness, amounting to a state of stupor, increase, and the animals die in a hectic state.

The treatment of rot in sheep requires the early removal to sound pasture; feeding on corn, peas, beans, and other nutritious grain ; allowing full doses of common salt and sulphate of iron in the food, and, when necessary, administering a purgative so as to keep the digestive organs in good order.

PARASITIC DISEASE OF LUNGS IN CALVES AND LAMBS.—
PHTHISIS PULMONALIS VERMINALIS, LUNGEN-WURMSEUCHE.

Next to rot, this is by far the most destructive disease of young sheep in the south of England. It is not so destructive in Scotland, but has injured farmers much this season in Ireland.[*]

If the lungs of sheep are examined in butchers' shops, a very large number of them will be found studded with deposits, once regarded as tubercular.[†]

This tubercle, in reality, consists in a deposit of ova of the *strongylus filaria* (Reed), surrounded by epithelium and granule cells, oily and crystalline deposit, with debris of healthy lung tissue. Generally this opaque and semi-gelatinous material is observed towards the more healthy part of the lungs in the shape of circumscribed masses, often not exceeding the size of an ordinary pin's head, and if each little nodule be squeezed, a gritty substance, the result of cretifaction of the above-mentioned deposit, is felt between the fingers. Each nodule indicates a spot where the germs of the strongylus filaria have been deposited, giving rise to irritation and the exudation of material around them; in this material granule and pus cells develope, and fatty, and lastly calcareous

---

[*] As an indication of the importance of this disease, I may mention that the farmers of Cornwall, through the Bath and West of England Society, recently offered a prize of £30 for an essay on this disease, which has been awarded to Dr Edward Crisp, who proves that the disease is due to overstocking, and especially to the feeding off a second crop of clover with lambs after the first crop has been consumed by sheep.

[†] I was not aware myself of the real nature of this deposit until 1854, when I had the privilege of prosecuting, with Dr Ercolani, of the Turin Veterinary School, some researches as to the methods of propagation of parasitic worms.

degeneration ensue. The eggs are of an oval shape. They are at first transparent, but in all those that are fecundated the yolk cleaves, and, by progressive subdivision of cells formed out of the yolk, a cellular mass is formed, which assumes an elongated and coiled appearance, and presents the external form and internal organisation of the strongylus filaria. The parasite, coiled on itself and alive in the cell, moves about, and at last becomes free and grows to its full size, passing out of the tissue of the lung into the air passages, whence it is coughed out and often deposited on grass and other substances likely to be eaten by the sheep. How it attains the lungs to deposit its eggs is involved in mystery, —perhaps by directly piercing the tissues from the stomach to the lungs; though, from the eggs being universally disseminated over the lung, we might be led to conjecture that the ova are introduced into the circulation and stopped in the pulmonary capillaries, where they produce irritation, and the deposit, before described, accumulates around.

The strongylus filaria is a worm from one to two and a half inches in length, the male smaller than the female and yellowish, whereas the latter is white. The body is of uniform size, but tapered at the extreme ends. Anteriorly is the head, short, stumpy, and matted, not tuberculated as that of other strongyli, but rather angular. From the mouth extends a short œsophagus into a short but elongated stomach, and from this the straight intestine extends back nearly to the extreme end of the tail, a little anteriorly to which is the anus. In the male an undivided circular aliform expansion, obliquely situated to the line of the body, surrounds a space in which the penis is observed. The tail of the female is pointed, the vulva situated near the anus, and from the vulva extend the oviducts full of eggs, and containing also live young.

In calves similar parasites abound under certain circumstances in the respiratory organs. The *strongylus vitulorum* (Reed), or *Str. micrurus* (Mehlis), is one of the armed strongyli with a filiform body, short caudal, long in the male, and mouth with three papillæ. This species is met with in the air passages of calves, and occasionally in the ass. Nicholls, in the first volume of the Philosophical Transactions, mentions the *husk*, common amongst calves under one year old, as dependent on worms in the windpipe; and in 1788, when Camper was engaged in investigating the cattle plague, and especially the advantages of inoculation as a preventive, he learned that one of his neighbours who had saved 50 calves by inoculation, lost 30 by this parasitic affection. On the 2nd of September of the same year, Camper had occasion to examine the trachea and lung of a calf that had died, as he expresses himself, with myriads of these worms in the air passages. On another calf Camper noticed a perfect ball of these worms effectually obstructing the windpipe. He described the worms well, and observed that they were viviparous. In his literary researches on the subject he found that Gesner had called a worm *Wasserkalb*, calf of water, of which he knew not the origin, but that calves swallowed them with the water to the great peril of their lives—*magno etiam vitæ periculo*.

In the pig a similar affection has been observed, and the worm has been described best by Mehlis and Gurlt. It has been called *strongylus paradoxus* (Mehlis); *gordius pulmonali apri* (Ebel); *ascaris filiformis cauda rotundata* (Goeze); *asc. bronchiorum suis* (Modser); and *strongylus suis* by Rudolphi, who looked on it as a doubtful species, having seen but two specimens which he had received from Bremser, and which had been found in the air passages of the domestic pig. Gurlt speaks of them as infesting the wild boar and the domestic pig, but that it is rare. Alessandrini, on the other hand, says that in Bologna he has found large numbers in the lungs of pigs killed in the public slaughter-houses, and it has since been recognised as frequent in Switzerland and France. The *strongylus paradoxus* has a narrow mouth, furnished with three papillæ; the caudal bag is bi-lobed, and turned downwards. In the female there is an enlargement where the anus is observed; the tail is short and pointed. The male is from eight to nine lines in length, and the female about an inch and a half. The females are by far the most numerous of the two.

Returning now to the parasitic disease of the lungs of sheep, it is clear that there are two distinct stages of the affection, the one mistaken for true tubercular disease, and the other when the worms are fully developed, and lodged in the air passages. Waldinger* was probably the first to give a good account of the latter stage, but the nature of the first was not brought to light until 1840, when La Harpe, of Lausanne, examined the affected lungs, and discovered the ova and young worms in the solid deposit, and recognised them as analogous to the strongylus filaria, met when full grown in the air passages, and sometimes in the act of piercing from the lung tissue through the mucous membrane into the bronchia.†

Unaware of La Harpe's discovery, Dr Ercolani, in 1843, when prosector to Professor Alessandrini in the University of Bologna, was struck with the appearance presented by

---

* Abpaulung äberd Wurmer und Lungen a Staape. Wien, 1818

† I do not agree with Dr Crisp's theory of the germs of the parasite being carried back from the stomach to the mouth in the act of rumination, and then finding their way into the trachea. As with the germs of the cysticerci only the young animals are affected, because they cannot pierce the tissues of older ones.

some sheeps' lungs he had purchased on the butcher's stall.
Many strongyli existed in the bronchia, and grey nodules or
tubercles on the surface of the lungs. These nodules
Ercolani found to contain small worms and eggs, in which
the young strongyli were in a state of development, already
alive and active. Since then Ercolani has made some interesting observations on the tenacity of life of the young
strongyli. These parasites show signs of life on being
moistened after drying for thirty days, and at other times
after having been immersed in spirits of wine at 30°, or in a
solution of alum and corrosive sublimate. Ercolani, moreover, says that the ova, abundant in the mucus of the bronchial tubes, containing worms, sink into the air vesicles,
become coated by an albuminous material, and thus are
imbedded in the lung tissue. This would lead us to believe
that when worms are swallowed by healthy sheep they immediately find their way into the windpipe. I must confess
I doubt this. Of course the eggs of the worms developed in
the lungs are deposited in the lungs again, or may move indirectly into the system of another animal, but the migration
from the mouth or alimentary canal to the lungs, certainly
requires a more complete explanation than has hitherto been
given. The number of embryo worms met with in the lungs
of one sheep is sufficient to infect a whole flock, so that the
disease has manifested itself as enzootic or epizootic.

Perhaps as early as La Harpe and Ercolani, did Dr C.
Radcliff Hall, of Torquay, investigate the question. In 1856,
in the British and Foreign Medico-Chirurgical Review, Dr
Hall says, "For fifteen years past I have been in the habit
of noticing the lungs in butchers' shops and slaughter-houses.
I have never seen a single specimen of the lung of a full-
grown sheep that was entirely free from entozoic disease.
The disease is not hereditary, since the lungs in young lambs

are healthy. Nor, I conclude, is it restricted to any specific locality, since I have found it at every place in Great Britain, France, Germany, and Switzerland that I have happened to visit. The lungs, then, of any full-grown sheep, taken indiscriminately, will be found to contain, and often to be thickly studded with, small nodules, varying in size from a pin's head to a barleycorn, or larger. The cysts are full of clear fluid, and contain cysticerci hanging upon an epithelial lining membrane. The firm, soft deposits consist of granule cells and molecular matter, in which minute ascaris like worms are found. The gritty nodule is one or other of these, which has undergone calcareous transformation. The particular point bearing upon my subject is, that the pulmonic affection does not prevent the sheep from furnishing excellent mutton." Further on Mr Hall has introduced a diagram to show the changes undergoing around the germs of the strongylus in the lung tissue, and says that there is nothing during the lifetime of the sheep to lead us to infer that it suffers pain, distress, or constitutional disturbance during the formation of this boundary of plastic inflammation around the nodules in its lungs.

Dr Ranke exhibited at the Pathological Society, on Tuesday, November 3, 1857, the lungs of three sheep affected with the disease, and he carefully described the morbid changes due to the parasites.

On examining the sheep slaughtered, we find that the larger number of them are fat and robust, yielding wholesome meat, but there is likewise a per-centage, and not a small one, conveyed to the butcher, because feeding cannot improve them, and to allow them time would be to allow them time to die by the disease. That the development of the germs in the lungs is always unattended with the slightest inconvenience, is not the case, and though the worms may not have found their way into the air passages, the changes going on in the early stages of the disease are associated with symptoms of spasmodic cough, irritation in the throat, and occasionally, as some of the small worms get free and coughed up into the nasal chambers, the sheep may be seen rubbing their head and nostrils on the ground, and sniffling to remove the cause of the irritation. It rarely happens, I believe, that large accumulations of worms in the lungs do not lead to emaciation, anæmia, and defective nutrition, with great debility and dropsy, unless the animals are suffocated by a lump of worms closing the windpipe.

It has been thought by some that the constitutional condition must precede the deposition of the germs and the development of the strongyli in the respiratory organs, but that this is not correct is proved by the animals continuing to thrive until, by the number of full-grown worms, the breathing is disturbed, the sheep are tormented, and fall back in condition. Other parasites accumulate in the liver or in the alimentary canal, and the animal falls into a state of hectic, with a manifest tendency to dropsy.

Concerning the prevention and treatment of this disease, it is only necessary to indicate, in the first place, the dangers attending the feeding of young sheep on the second and third crops of clover, after the first has been fed off by older sheep. To prevent the disease, you require fresh and sound pasture, and it may be necessary to supply a considerable quantity of artificial food. To cure the disease, inhalations of chlorine gas are recommended, or the internal administration of camphor and turpentine, in oil or ether. Sound food, such as oats, linseed-cake, cotton-cake, turnips, &c., must be allowed, with ferruginous tonics. The iron may be given to the extent of ten or twenty grains daily to each lamb, with a drachm of common salt.

ON PENTASTOMA TÆNIOIDES OF THE SHEEP.

Leuckart* has recently shown by experiment that Pentastomum denticulatum, which is found not unfrequently in the bodies of rabbits, is the partially developed Pentastoma tæniiodes which occurs frequently in the nasal cavities of the head or sinuses of the dog. Moreover, Leuckart has shown, that when this parasite has attained maturity in the dog's head, ripe eggs are thrown off to ensure the multiplication of the species. These eggs are given off and discharged with the mucus in the act of sneezing, &c., and they are then taken up by animals, in whose bodies the embryos undergo a certain stage of development.

The pentastomum had not been seen in its undeveloped state as scolex in all our domestic animals. It had been found in the abdominal cavity of goats and cats. In the first, the parasite was the pentastomum denticulatum, and in the second, pentastomum fera. Colin, however, has recently found it in the sheep and dromedary.†

In the mesenteric glands of the last-named animals, there exists an asexual linguatula, which acquires a generative apparatus on changing its habitation. These parasites penetrate the gland, and are lodged in a capsule which contains several individuals. As the containing capsule enlarges, disease of the gland tissue occurs. The parasite of the mesenteric glands is born of the eggs of the parasite of the dog, which are gathered up by the sheep with their food. The worm only remains a definite time in its first abode, as it pierces the glands and leaves a cavity which soon gets filled up.

---

* *Bau und Entwickelung der Pentastomen.* Von R. LEUCKART. Leipzig and Heidelberg. 1860.

† See *Edinburgh Veterinary Review*, vol. iii. p. 682.

When a dog or wolf eats the entrails of animals in whose glands the parasites exist, the embryo may adhere to the lips and nose, and then pass into the nasal cavities. Fürstenberg says that the linguatulæ pass up the nose rapidly, and fix themselves by the hooks so as not to be expelled in the act of sneezing. These worms, which so suddenly change their habitat, increase in size, and their generative organs are developed in less than two months. They must remain a year in the nose of the dog, in order to attain complete development.

There can scarcely be a doubt, says Fürstenberg, that the linguatulæ found in mesenteric glands of sheep, belong to the same species as those discovered in cysts in the lungs of rabbits, and whose complete development in the dog Leuckart has witnessed.

Fürstenberg has therefore confirmed Colin's observations, and added some new facts as to the escape of the linguatulæ from the mesenteric glands.

EPIZOOTIC AND ENZOOTIC DISEASES OF THE HORSE.

PERIODIC OPHTHALMIA.

This is an affection of the eyes peculiar to the equine race, and which is often incurable and eminently destructive to the organs affected. It is a constitutional disease, localizing itself in the eyes, and generally leading to blindness. To this malady the names of *periodic* or *specific ophthalmia* are given, on account of the certainty of its recurrence even after an apparent cure; and that of *moon blindness,* from its recurring monthly, or, as was supposed, with special changes of the moon. It is sometimes sporadic, but at others enzootic, and affects a large proportion of the animals in a district. A knowledge of the causes of the malady is thus of much more importance than that of any system of treatment that has been suggested. The causes are *predisposing* and *exciting*. Among the predisposing causes may be mentioned the following:—

1*st*. Soils of a clayey and humid character have a deleterious influence on horses raised on them. These are soft and flabby, with a predominance of areolar tissue, thick skins, long hair, flat feet, and a general lymphatic temperament. Such horses seem more susceptible to morbid influences, and especially to those producing this disease. If removed to dry calcareous soils, the predisposition may remain latent throughout life; but, on the other hand, horses bred on the latter soils, and afterwards removed to the former, are very liable to contract the malady.

2*nd*. Soils naturally damp from their own character, or from that of the subsoil, and which have not been ameliorated by drainage, have the same influence on the constitution of the horse, and predispose to this affection in the same way as the argillaceous.

3*rd*. Excessive humidity of the atmosphere exerts a similar determining tendency, and, accordingly, periodic ophthalmia is a common disease on the banks of large rivers and lakes, and, in some cases, in the vicinity of the sea. So great is the influence of the peculiarities of soil and climate on the development of this disease, that Spanish dealers are said to buy up affected animals in the south-western departments of France, being convinced that, if they are transported to certain regions beyond the Pyrennees, the malady will disappear. Many parts of Ireland afford excellent examples of the truth of the above remarks.

4*th*. Fodder of inferior quality, which contains a larger amount of aqueous and few nutritive principles, has a debilitating effect on the general system, and thus predisposes to the disease. This influence is especially marked in the case of foals early separated from their dams and supported on fodder raised on marshy pastures. Horses imperfectly supported in other respects are similarly predisposed.

5th. Certain kinds of eye seem especially disposed to contract this malady. It is more prevalent in the small sunken eye than in that which is full, bright, and prominent. Percivall mentions black eyes as being most obnoxious to the malady, while Messrs Castley and Goodwin have seen lighter eyes suffer in an equal degree.

6th. Consanguinity is mentioned by Reynal as a predisposing cause, its mode of action being by reducing the stamina of the progeny.

7th. Of all influences tending to the development of the disease, none is more clearly established than the hereditary predisposition. Our best Yorkshire breeders would not employ a blind sire or dam, and, where such are had resort to, the progeny usually inherit the propensity in a marked degree. A similar conclusion has been arrived at by the best English and foreign veterinarians; and Reynal remarks that predisposition may remain latent in one generation and re-appear in the next.

Young horses, about the time of teething, are frequent subjects of the disease, and, accordingly, the French and some English veterinarians attribute it to the local plethora attendant on dentition. Percivall and D'Arboval have noticed its greater prevalence in geldings than in mares, a circumstance which has been explained by the greater local irritation in the former, connected with the cutting of the canine teeth.

Symptoms.—The malady may make its onset slowly, and show itself by a profuse flow of tears and some redness of the conjunctiva; but more commonly it originates suddenly, often during the night, and is first recognised by swollen eyelids, nearly closed, abundant lachrymal secretion, strongly injected conjunctiva, opacity of the cornea, protrusion of the haw over the globe of the eye, and considerable intolerance of light. It is generally referred to a blow; but the evident pain on exposure to light will often show that the deeper parts of the eye are involved. There is some fever indicated by hot dry mouth, hard pulse, and slight costiveness, but the appetite may still be good. On the second, or from that to the sixth day, the opacity of the transparent cornea becomes more marked, the whiteness being referrable to the interior of the eye, and a close examination shows this to be composed of a number of albuminous flocculi floating in the aqueous humour, and entirely hiding the iris. These flocculi have a white, or dirty yellowish-white, colour, and, in a day or two after their appearance, become in great part deposited in the lower part of the anterior chamber. They subsequently change to a greenish or brownish hue.

The symptoms commence to disappear at a time varying from the fourth to the tenth day or even later, and the tenderness becomes gradually removed, the intolerance of light ceases, and the cornea becomes clear. In the turbid aqueous humour numerous flocculi remain, and behind it the iris is seen, with a dull greenish or brownish aspect. As the absorption goes on, the cornea and anterior chamber become perfectly clear. The duration of an attack may vary from four or five days to forty. The first attacks are usually the longest, and their duration diminishes, as a rule, with their recurrence. During the progress of apparent recovery a relapse is not unfrequent, and the term may be thus indefinitely lengthened. The interval between the attacks is, on an average, about sixty days. The eye may seem quite clear during the intermission; but it has not returned to its normal condition. The outline of the upper eyelid is usually altered. It presents a slight bend in its internal part, so that the upper joins the lower lid, at the inner angle, by a right in place of an acute angle. This is best marked after several severe attacks, and gives a triangular outline to the opening between the lids. The iris of the affected eye is more contracted than the opposite; it has lost its lustre, and does not contract and dilate to the same extent as in the normal state, on the access of light and darkness. With the aid of the opthalmoscope, fibrinous deposits can commonly be seen on its surface, delicate flocculi in the anterior chamber and on the anterior aspect of the lens, and the choroid is observed to be altered in hue, with slight elevations on its surface.

The common termination of periodic ophthalmia is in cataract or opacity of the crystalline lens or its capsule. The iris sometimes gets attached to the anterior aspect of the capsule of the lens, and either remains permanently fixed, or, becoming lacerated during its movements, retains for the future a ragged margin, while the adherence of the colouring matter to the anterior aspect of the lens constitutes lenticular cataract. Amongst the ultimate results of this malady, Reynal enumerates opacity of the cornea; slight turbidity of the aqueous humour, with greater thickness and a more glutinous character; the formation of false membranes on the iris; metamorphosis of the lens, more or less completely, into a fibrous or cretaceous structure; adhesion of the lens to its capsule; partial or complete disappearance of the lens, probably by absorption; adventitious deposits in the vitreous humour, especially at its posterior part, where, in bad cases, there may be calcareous deposit; fibrinous deposits in the retina, and atrophy of the optic nerve as far as the corpora quadrigemina; similar deposits in the choroid, which presents numerous small rounded eminences; and, lastly, similar products of inflammation on the inner aspect of the sclerotic, and puckering of its substance commensurate with the absorption of the liquid contents of the eye.

Treatment.—In a disease such as that before us, preservative measures are much more effective than curative. These will consist in counteracting the various causes which predispose to the malady. Among them, efficient drainage, ventilation, and cleanliness; the use of aliments of good quality, and the avoidance of sires or dams that may have suffered from the affection, will hold prominent places.

Its therapeutical treatment is unsatisfactory. Bleeding—local and general—scarifications of the conjunctiva, purgation, fomentations, blistering, setons, rowels, collyria—sedative, astringent, and caustic—and numerous medicaments given internally, have been resorted to in turn, with only partial success. With or without these measures, the disease sooner or later disappears; but it is only for a time; and if one or other of the agents enumerated hastens the ameliorative process, very little has been gained, no medicinal measures can secure the eye against a succeeding series of attacks, until the inevitable consequence, the destruction of vision, results. Different veterinarians have recommended the application to the eye of strong solutions of nitrate of silver, bichloride of mercury, and the like; though perhaps an equal amount of good will accrue from the employment of milder astringent lotions of similar agents. Belladonna may be employed in the form of a lotion, to be applied to the eye daily during the severity of an attack, the object being to dilate the pupil, and break up any connections that may form between the iris and lens. The ordeal bean of Old Calabar has been employed in human medicine to contract the pupil, and, being applied alternately with belladonna, will move the iris through a greater space. When with the fever there is costiveness, a mild laxative will prove beneficial. In all cases, darkness will be found not only grateful to the patient, but also highly favourable to recovery.

INFLUENZA.

The term *influenza*—the Italian for INFLUENCE—is a highly improper one, but it is applied to a catarrhal or rheumatic affection usually associated with much derangement of the liver, subacute or latent inflammation of the pleuræ, and a low form of fever commonly called *typhous* or *typhoid*. The marked feature of the disease is its appearance amongst many animals simultaneously, often laying up all the horses on a farm, and carrying off several. It is especially within the last thirty years that information has been collected concerning this occasionally very prevalent disease, and its supposed—but not proved—greater frequency of late years, has been regarded as evidence of a change of type of disease from that form in which animals would bear free blood-letting to certain kinds in which active bleeding or purging are usually followed by great weakness and often by death. We can trace back to 1819 distinct outbreaks of the same influenza that we witness occasionally at the present day.

Influenza is essentially a protean malady, varying in its source in different outbreaks, and much according to the state of the weather. It is most common about spring and autumn, and said to be most dangerous during the prevalence of easterly winds. Influenza attacks young horses more than old, and is occasionally communicated from the sick to the healthy by infection or contagion. Veterinary surgeons have noticed that the animals they have ridden about in their practice have caught the disease from patients they have been attending, and there is usually such a succession of cases when one marked instance of influenza occurs, that it is likely that contagion exerts some influence in the spread of the disease. Continental authors, such as Professor Hering, describe three forms of influenza:—1st, The catarrho-rheumatic form ; 2nd, The gastric, or bilious rheumatic form; and, 3rd, the gastro-erysipelatous form. In this country the line of demarcation between different forms of influenza has not been so finely drawn, though that there are differences may be gleaned from the recorded notes on various outbreaks. As a rule, with symptoms more or less prominent of a general febrile condition, there is great dulness and debility, frequent and weak pulse, scanty discharge of dry excrement and high-coloured urine, appetite lost, and there are often decided signs of jaundice. The eyes are more or less sunken, upper lid drooping, the conjunctiva of a yellowish-red colour ; the buccal membrane of a similar tint, and the lips hanging; the animal's skin is dry and coat unhealthy-looking. In many cases of influenza, cough, and sore throat, a tendency to catarrh, or, in more common cases, to subacute or latent pleurisy, are characteristic features of the disease. In other cases, again, the symptoms of stomach staggers and partial paralysis of the hind quarters occur. In this, as in other forms of the disease, there is a great disposition to œdematous swellings, which have been regarded as erysipelatous on the Continent.

In 1861 a somewhat general outbreak of influenza occurred, which Mr Chapman of Gainsborough, in Lincolnshire, describes as having proved a fearful disease, " commencing in the *first stage* with *great prostration*, excited respiration, in some cases terribly laboured, accompanied with a painful grunt, ending in a sigh; pallid membranes ; mouth filled with frothy mucus, very fœtid ; extremities deadly cold ; pulse, in many of the worst cases, imperceptible at the jaw, in the milder attacks very feeble, running up to 70 or 80.

In some cases there *seemed* abdominal irritation; the animal crouching, and pawing occasionally with the off fore foot; bowels sluggish in action, but *not* constipated. Auscultation showed *partial* congestion in the lung or lungs, with feeble and *peculiar action* of the heart."

The *post-mortem* appearances of influenza vary according to the character of the disease and the complications which arise during its progress. Very generally there is effusion on the thorax, recent adhesions, evidence of pericarditis, and sometimes of inflammation of the pericardium. In all forms of influenza of a rheumatic type, the lesions just named are discovered; whereas, in others, there is more evidence of bronchial catarrh, pneumonia, and the blood-changes are indicated by ecchymoses beneath the serous membranes, especially in the cardiac surfaces.

The treatment of influenza consists in placing the animal where it can breathe fresh air, and be without restraint. A loose box is the best place to keep it in. A mild purge I find usually to benefit the animal, using Cape instead of Barbadoes aloes. This is followed up by the following:—

Nitrate of potash . . .	2 oz.
Carbonate of ammonia . .	1 oz.
Solution of the acetate of ammonia .	12 oz.
Water	12 oz.

This is to be divided in four doses, to be given night and morning. If effusion in the chest is threatened, I prefer giving digitalis and nitre in diuretic doses thrice daily for a couple of days; and the benefits derived from this treatment are great. Mustard poultices may be applied to the chest, but I do not approve of active blisters, rowels, or setons. When the acute symptoms subside, ferruginous tonics are highly beneficial, and the animals require a liberal diet and moderate exercise. Blood-letting, and other active depletive measures, cannot be too strongly condemned in the treatment of this disease.

FARCY AND GLANDERS.

Under certain circumstances a specific disease is developed in the horse, characterized by the formation of a virulent animal poison, which is very destructive when introduced into the system of man and various warm-blooded animals. This specific disease is termed *farcy* when the local manifestations affect the skin; whereas, in a more severe form, the lungs are implicated, and the system is so generally impregnated with the poison as to destroy life with certainty, sooner or later—it is then called *glanders*. The terms farcy and glanders are meaningless and inappropriate. The first is an importation from the French by Sir William Hope, who translated Solleysell's work on the horse, and reproduced in the English version of this work the French word *farcin*. The second is derived from the word *glans*, gland, and has been applied to the disease from the enlargement of the submaxillary lymphatic glands met with in every case of this disease.

Farcy and glanders are diseases of temperate climates. They are unknown in very hot countries, and rare in very cold ones. They are capable of spontaneous origin in the horse, and especially is this the case with farcy; but the common cause of both affections is contagion. It is extremely rare to see a case of farcy or glanders that we cannot trace to communication from the diseased to the healthy; but in very foul stables, in badly ventilated mines, in the holds of ships on long and rough sea voyages, glanders may break out with great virulence. The two diseases are rife in times

of war amongst horses engaged in the transport service, as
well as those used by the cavalry. The prevalence of the
diseases, under these circumstances, has been ascribed to
privation, dirt, and bad management. Probably it is not a
little due to the necessary congregation of animals of all
kinds, healthy and unhealthy, and to the facilities for inoculation amongst animals often bruised, cut, and otherwise
injured. It has been noticed, that mules engaged in the
transport service in times of war suffer very severely from
acute glanders, and, like asses, they are rarely attacked by
chronic farcy or the slow and insidious form of glanders. In
the French army the average annual loss by glanders amounts
to 23.8 per 1000 horses, and 1.5 per 1000 die of farcy.

Symptoms of Farcy.—In the acute form of this disease,
there are first the symptoms of irritative fever, such as
shivering, followed by heat of body, frequent pulse, hurried
breathing, dulness, &c. In the chronic form, the local signs
first appear, and they consist in papulæ, or circumscribed
inflammatory swellings of the skin, either situated around a
wound, or on the course of the principal vessels of the head,
trunk, or limbs. One or more isolated swellings or buds
may occur, but usually they are seen to congregate, and are
more or less connected with each other by corded lymphatics, which are very tender on pressure. Generally in the
course of these inflamed lymphatics, slight tumefactions occur
at the seat of the valves, and wherever a swelling begins a
farcy bud forms. Each farcy bud suppurates; the pus discharged may at first be laudable, but is soon more or
less ichorous and irritating. The opening through which
each abscess bursts becomes enlarged by ulceration, and has
no tendency to heal. The part which is the seat of the
eruption may be more or less diffusely inflamed, and this
happens especially with the hind extremities, which are the

most frequently affected. The fore limbs suffer next in point of frequency, and next to the extremities we find the head most commonly implicated. The lymphatic glands in the vicinity of the part affected tumefy. Farcy may kill in eight or ten days, or continue in a chronic form for several months. When it kills it has usually been followed by glanders.

Symptoms of Glanders.—As in farcy, we find the premonitory signs of fever in this form of disease, and the animal dull and dispirited, with discharge at the nose. The discharge is at first watery and then purulent ; it may attack only one nostril, right or left, or both, and the submaxillary lymphatic glands are swollen, solid, and have a tendency to adhere firmly to the inside of the jaw. The nostrils are often swollen, and more or less closed by the glutinous discharge, which soon becomes fetid and sometimes sanious. On opening the nostrils, pustules and ulcers are seen on the Schneiderian membrane, and as the disease advances, the ulceration extends so as even to lead to an open communication between the two nasal chambers. In some mysterious cases of glanders, the ulceration occurs in the false nostril ; and I have often seen French veterinarians pass their thumb into the false nostril to feel for any such ulcer.

Ulceration of the Schneiderian membrane may occur in other diseases of the horse, but so rarely that we are apt to look upon it as truly diagnostic of glanders, and the diagnosis is confirmed when, on auscultation, the lungs are found implicated. The signs of the lung complications are,—difficult breathing ; considerable constitutional disturbance in many instances, whereas in others there is simply dulness and percussion ; and absence of respiratory murmur over considerable portions of lung, with tolerable resonance and loud murmur over others. The lung affection consists in the development of many abscesses interspersed throughout the lung tissue, and often associated with some amount of pneumonia. Too little attention has been paid to the auscultatory phenomena of cases of glanders, whether acute or chronic. When the lungs become much affected, the animal's appetite is disturbed, there is considerable weakness, anæmia, and tendency to wasting.

Death from glanders is due, in many instances, to the direct effects of blood-poisoning, characterized by stupor, fetid secretions, and impeded circulation. In other cases the animals sink in a state of hectic.

Farcy may be successfully treated by applying blisters round the buds, and caustics to the ulcers, whilst the system is supported by tonics. The remedies employed with greatest success internally have been arsenic, cantharides, sulphate of copper, sulphate of iron, quinine, and other vegetable and mineral tonics. Locally, the free use of antiseptics, such as chlorine water, carbolic acid, Condy's permanganates, &c., is to be recommended. Where an eruption of farcy buds occurs around a fetlock joint, and in the vicinity of a wound which is much inflamed, a poultice may be applied composed of linseed meal. When the warm poultice is made, a few ounces of liquor plumbi diacetatis can be mixed with it with advantage, and caustics afterwards applied. Attention must be paid to the organs of secretion, and laxatives or diuretics are required at intervals.

Glanders is an incurable affection, and any animal suffering from it should be instantly destroyed. We cannot too strongly condemn the practice of experimenting with glandered horses, as the danger of infecting men and animals is too great to be trifled with, and prevention is decidedly better than cure under such circumstances.

CHAPTER XVII.

NERVOUS ACTION.

Muscular irritability and nervous stimuli.—Reflex actions.—Physical nervous actions.—Mental nervous actions.—Nerve-cells and fibres.— Their functions.—Analysis of nervous matter.—Spinal cord.—Its structure and functions.—Functions of the groups of cells.—Crossing of sensitive impressions on the cord.—The brain.—Oblong medulla.— Pons varolii.—Cerebrum.—Cerebellum.—Their functions.—Uses of the medulla in respiration and deglutition.—Vital point.—Effects of pricking the floor of the fourth ventricle.—Effect of removing the cerebellum.—Turning or rolling as a result of injury to certain parts.— Removal of the cerebrum in birds, and its effects.—Cranial nerves.— Olfactory nerves.—Optic nerves.—Their functions.—The influence of their crucial and other fibres.—Auditory nerves.—Common motor nerves of the eye.—Pathetic nerve.—Abducent nerves.—Trifacial nerves.— Their varied functions.—Sympathetic ganglia on their course.—Facial nerve.—Its influence on the salivary glands.—Glosso-pharyngeal nerves —Their influence on deglutition.—Pneumogastric nerves.—Chauveau's experiments.—Effects on breathing.—Effects on the lungs when divided.—Action on the heart; on the stomach.—Spinal accessory nerves; their influence on the voice.—Hypoglossal nerves.

THE nervous system is that part of the animal body to which all others are subservient, since it possesses the power of controlling and harmonizing the various functions essential to life.

The organs and tissues of the economy are, by virtue of properties inherent to them, capable of responding to stimuli or exciting causes, in such a manner as leads to the exercise of their individual functions. This inherent property is

known as the excitability or *irritability* of the tissue. It
is well exemplified in the contraction of muscular tissue
on the application of a mechanical or chemical stimulus.
Thus, if the point of a knife be applied to the muscle of a
recently slaughtered animal, a contraction is induced of a
more or less powerful character, according to the previous
healthy and vigorous condition of the muscle. Again, if
the leg of a frog be separated from the thigh, the skin
removed, and the poles of a galvanic battery brought into
contact with the surface of the exposed muscles, an ener-
getic contraction takes place whenever the electric circuit
is completed. In the healthy condition, this excitability is
called into play through the nervous system, so that all the
animal functions, whether these be of nutrition, sensation,
secretion, absorption, or locomotion, &c., are subservient to
this important apparatus.

In this case, however, the stimulus is not usually applied
directly to the part the functions of which are to be acti-
vated; it is more commonly originated at a remote part of
the system, or in the brain as the consequence of a mental
act, and, in either case, is conveyed through the nervous
system to the organ it is destined to affect. In this manner
it is that the contact of sapid substances with the mucous
membrane of the mouth leads to a secretion of saliva, or
that the falling of luminous rays on the delicate expansion
of the nerve of sight induces a contraction of the pupil.

It will be further noticed that these reflex actions, as
they are called, when the impression is received at one
part, and conveyed by the nervous apparatus to an organ
more or less remote, co-operate to bring about the healthy
exercise of the vital functions, and the maintenance of the
frame in a condition of integrity. This is well exemplified
in the above-named instances. The secretion of saliva,

when aliment has been introduced into the mouth, is necessary to assist in mastication, deglutition, &c.; and when a powerful light falls upon the eye, the closing of the pupil is no less needful to intercept the luminous rays, and to prevent them striking on the bottom of the eye, dazzling the sight, and otherwise injuriously affecting the organ. The acts above cited are known as *physical nervous actions*, since they take their origin in a physical impression on some part of the economy. By means of this mechanism are sustained all the important vital functions, such as circulation, respiration, digestion, &c. In all such cases, the stimulus must be properly regulated as regards character, force, time, and direction, as otherwise it might prove not only abortive but even eminently injurious.

In *physical nervous actions*, the primary impression and the resulting movement may, one or both, be taken cognisance of by the mind, or *vice versa*. Thus, when a limb is suddenly withdrawn as a result of contact of the toes with a hot or other irritating body, both impression and result are quite patent to the mind, though the muscles of the leg were in action before the will had time to command. The act of breathing, of which we are quite conscious, results from an impression made by the circulating blood on the system, but especially on the lungs and brain, and of which we are unconscious, unless we for a time voluntarily suspend the respiratory process. Lastly, as an example of the exercise of both without the mind perceiving it, may be mentioned the movement of the intestines, as a result of contact between the ingesta and the mucous membrane. It may be concluded, that though the mind may take cognisance of, and even perfect such an action commenced independently of it, still in no case is perception or the exercise of the will necessary to the performance of such an act.

In addition to those spoken of above, may be mentioned physical nervous actions due to a morbid origin. Such are the involuntary movements which sometimes take place as a result of diseased conditions of the great nervous centres. These differ from those already noticed in not being due to reflex action, and accordingly bear an analogy to mental nervous actions.

Besides the purely physical nervous acts, there is another class of necessity connected with the mind, and hence termed *mental* or *psychical nervous actions.* These are of three kinds—*acts of perception, of emotion,* and *of volition.* 1st, Acts of perception include general and special sensibility, &c.; an impression in this case is made on some part of the body, and from that conveyed to the mind. 2d, Acts of volition, in which any portion of the body is moved in obedience to a mandate of the will: this, unlike the last, originates with the mind, from which the necessary stimulus is conveyed through the nervous system. 3d, Acts of emotion, with which we shall have less to do, are those actions which originate in peculiar psychical conditions, as in joy, anger, fear, &c., and which are in great part independent of the will.

In connection with the above mentioned functions, it is worthy of note, that, of all the systems of the animal body, the nervous alone is that upon which the mind can directly act, or which can act immediately upon the mind. This system, moreover, is peculiar to animals, and is a distinguishing characteristic of the latter, as compared with the vegetable creation. In the words of Dr Todd, "it is obviously the presence of a psychical agent, controlling and directing certain bodily acts of animals, which has called into existence the peculiar apparatus which the nervous matter is employed to form."

The nervous system is made up of two elementary structures—*cells* and *fibres*. "The nerve-cell has been described as consisting of an envelope, granular contents, and a nucleus with one or more nucleoli. So far, there is no difference between the so-called nerve-cell and any other cell; but it is the great diversity in size, its frequent coloration with pigment, and the brilliancy of a vesicular nucleus which may be regarded as somewhat characteristic of the nervous element, which is usually stellate, round, or

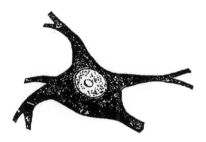

Fig. 208.—Multipolar nerve-cell, showing its nucleus and nucleolus.

oval, and connected with fibrous prolongations. Some of the cells are as large as $\frac{1}{300}$th of an inch in diameter, and they are not unfrequently as small as $\frac{1}{3000}$th. Vesicular bodies are interspersed amongst the cellular elements."

"The nerve-fibres are found in the nervous centres, and in the nerves which connect the latter with the peripheral parts of the body. They are combined with, or spring from, the nerve-cells in the grey matter of all nervous ganglia. The ultimate nerve-fibres are of two kinds—tubular or white and grey or gelatinous. The white have a *special envelope,* in the interior of which is Remak's primitive band or axis cylinder (Purkinje), surrounded by a medullary sheath or white substance of Schwann. The

tubular fibres have been called medullated by Kölliker, to
distinguish them from the non-medullated. The latter
occur in organs of special sense, when delicate plexuses
are formed as in the retina, or the olfactory and auditory
apparatuses, and have a structureless envelope containing a
clear granular axis without the white substance."

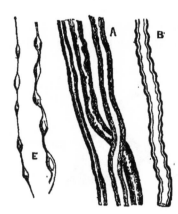

Fig. 209.—Tubular or white nerve-fibres. A indicates the dark outline produced in the
medullary sheath by exposure; B shows the double contours in a nerve-fibre; and E
the varicosities resulting from traction or pressure.

" Gelatinous fibres have been described by Remak, Henle,
and others, and are found associated with the elements
already noticed. They are flat, slightly granular, nucleated,
and about $\frac{1}{8000}$th of an inch in diameter."—*Anatomy of
the Domestic Animals. Gamgee and Law.*

The nerve-cells which exist in the grey or cineritious
matter of the brain and other nervous centres, are the gene-
rators of nervous force and recipients of sensory impressions, and their presence in any part indicates it to be a
nervous centre or ganglion. They generate nervous force,
however, only when a proper stimulus is conveyed to them,
as referred to under the heads of Physical and Mental

Nervous Acts. In some few instances they may be supposed to originate such force, as exemplified in the persistent contraction in which are maintained the sphincter muscles of the anus and neck of the bladder. Some nervous centres (in the brain) are alone capable of taking knowledge of sensory impressions; and these impressions accordingly, if from distant parts, and especially from such organs as are supplied by the sympathetic nerve, have to traverse one or more ganglia before they can be made patent to the mind.

The nerve-fibres are incapable of generating nervous force of themselves. Their sole functions are to convey such force generated in the nervous centres to the peripheral parts of the body, and to convey sensations and impressions from such parts to the centres. They are simply conductors of nervous force, and are aptly represented by the conducting wires of an electric machine, whilst the centre, as the generator of force, is the analogue of the machine itself.

The white fibres make up a great portion of the brain and spinal cord, as well as almost the entire bulk of the nervous trunks, with which these are directly connected. They seem much more tough and resistent in the nerves than in the centres; but as the fibres are in the two cases identical, this appearance is entirely dependent on the amount of white fibrous tissue by which the nervous fibres are enveloped.

The grey fibres make up the main part of the various branches of the sympathetic nerve, and enter in varying proportion into the formation of the cerebro-spinal nerves.

Nerve-fibres lie side by side in the nerves, but throughout their whole course they maintain an entire independence of each other. There is no union of individual fibres in the nerve, nor any bifurcation of single fibres to permit a more extensive distribution. A nerve accordingly possesses a

definite number of fibres at its point of origin from the
nervous centre, and has neither more nor less at its peripheral extremity, each fibre passing in an unmodified condition from its origin to that part of the tissue or organ in
which it is destined to ramify. Nerves, it is true, frequently
become connected or anastomose with each other, but this
union results alone from an interchange of fibres, while
each of these continues to maintain its distinct individuality.
The anastomoses of nerves is thus principally intended to
ensure a wider distribution of nervous fibres coming from
the same centre, and to obviate to some extent the occurrence of paralysis from injury to a single centre or nervous
trunk. Where a free anastomosis of this kind take place
between a number of nerves, the structure is known as a
plexus.

Regarding the origin of fibres in a nervous centre, there
is some difference of opinion. Some anatomists assert that
the fibres form loops which lie in the grey substance and
in contact with the nerve-cells, with which, however, they
have no direct structural connection. On the other hand
it has been clearly demonstrated that in the ganglia the
fibres arise directly from the tails of the caudate (stellate)
cells, and Schröder van der Kolk and others have satisfactorily shown that a similar origin is at least frequent in
the spinal cord.

The modes in which nervous fibres terminate in the
tissues are varied. The nerves commonly break up into
small branches, which arrange themselves in plexuses, and
from these individual fibres are given off to terminate as
follows : 1st, *In loops,* a single fibre bending backward in
the substance of the tissue and entering either the same
or an adjacent nervous trunk, in which it is understood to
follow a retrograde course to the nervous centre ; 2d, Some-

times the fibres seem to loose their dark outline and white substance, become less distinct, and are ultimately lost in the substance of the tissue; 3d, In some cases the ultimate nerve-fibres seem resolved into minute plexuses, as in some of the serous membranes; 4th, Sometimes they terminate in free ends which may enter minute ovoid bodies—*Pacinian bodies*—met with at certain parts of the surface of the true skin; 5th and lastly, they may terminate, as in the eye and ear, by becoming connected with true nerve-cells. The presence of such cells, however, constitutes the part a true nervous centre.

Nerve-fibres, we have said, are conductors of impressions and of nervous force. It is remarkable that in the performance of these different acts special fibres are employed, and this remark applies equally to the cerebro-spinal and sympathetic systems. Thus, in both alike there are certain fibres which conduct impressions only towards the nervous centre (*sensory, afferent,* or *centripetal*), and others conduct only *motor force* (*efferent* or *centrifugal*). No nerve fibre can convey more than one kind of impression. There is no difference in the size or appearance of these fibres to indicate their special functions, so that their relative properties can only be ascertained by observing their action under the influence of a stimulus. This is ordinarily supplied to sensory nerves by external objects applied to their extremities, and to motor by the will, or some reflex stimulus through the medium of a nervous centre. Any stimulus, however, applied to the trunk of a nerve, is sufficient to bring it into a state of activity; and it does not import whether this stimulus is mechanical, chemical, electrical, or a simple excess of heat or cold. The application of any such excitant to the trunk of a sensory nerve leads to the idea of pain, &c., not only in the irritated point, but also in those parts in which the peripheral ends of the nerve are situated. It is on this account that, after the amputation of a limb, pains are often endured, which are referred by the patient to the excised member. If, on the other hand, one of these stimuli be applied to the trunk of a motor nerve, a contraction takes place in the muscles to which it is distributed, notwithstanding that the nerve may have been divided in the interval between the part irritated and the nervous centre. It is thus seen that the functional activity of a nerve may be aroused by a simple modification in the condition of its fibres, and the effect is the same as if it had originated in the nervous centre or peripheral extremity as the case may be, and had been conveyed along the whole length of the constituent fibres.

Nerve-fibres can only act in one direction. Thus, if a sensory nerve be divided, and the peripheral extremity subjected to irritation, no sensation is experienced. Irritation of the portion still in connection with the brain, however, gives rise to the most lively pain, referable, as above stated, to the part from which the nerve conducted. In the same manner, if a motor nerve be divided and an irritant applied to the end still in connection with the brain, no effect either sensory or motor is produced; but when that in connection with the muscles is treated in a similar manner, a contraction in the latter immediately follows. These experiments can only be satisfactorily made by cutting down upon and irritating the inferior and superior roots of the spinal nerves, the former being exclusively motor, the latter exclusively sensory. At any other point of their course, the nerves contain fibres of both varieties, and no relative experiment can be made.

When a section has been made of a motor nerve, and its detached end stimulated so as to produce a powerful muscular contraction, cries of pain are frequently elicited, which Bernard considers due to what he styles a *recurrent sensibility* in the nerve. Brown-Séquard, however, and Chauveau, the able teacher of anatomy and physiology in the Lyons Veterinary College, who have carefully investigated the subject, deny the truth of Bernard's theory, and attribute the pain to the violent muscular contraction alone.

It is worthy of note, that if a nerve is subjected to a serious injury, it often loses for a time all power of performing its accustomed functions. Of this kind is the condition of shock which supervenes on serious accidents, and in which the patient may remain for hours quite unconscious of any painful sensation.

Chemically considered, nervous-tissue is composed of water, albumen, fatty matters, and salts. No perfectly satisfactory relative analysis of the white and grey nervous matter has been made. The best is the following, by Lassaigne :—

	Grey.	White.
Water, ...	85·2	73·0
Albuminous matter, .	7·5	9·9
Colourless fat, . .	1·0	18·9
Red fat, . . .	3·7	0·9
Osmazome and lactates,	1·4	1·0
Phosphates, . . .	1·2	1·3
	100·0	100·0

Fremy states that the fatty matters consist of cerebric acid, which is most abundant, oleic, margaric, and oleophosphoric acids, and of cholesterine. He remarks, further, that in the brain the fatty matters are confined to the white substance, and that the latter loses its colour when these have been removed. Vauquelin remarks that the spinal cord contains more fat than the brain ; and L'Heritier, that the nerves contain more albumen and soft fat than the brain.

In Mammalia, the nervous system is in two great sub-divisions—the *Cerebro-spinal* and the *Sympathetic.*

The *Cerebro-spinal* system, called by Bichat the nervous system of animal life, includes the brain and spinal cord, together with the nerves connected with these centres, and the ganglia situated on the course of the nerves or in the brain. This division presides over those acts with which the mind is more immediately connected,—as the mental, and even a large proportion of the truly physical nervous actions.

. The *Sympathetic* or *Ganglionic* system presides over those physical nervous actions which are . not directly connected with the mind,—as the functions of digestion, circulation, nutrition, &c. On this account it has been named by Bichat the nervous system of organic life. It consists of a series of nervous ganglia, lodged in different parts of the body, connected with each other, and, to a less extent, with the *Cerebro-spinal* system, by nervous cords, and sending off nervous trunks to ramify in the surrounding tissues.

ANATOMY AND PHYSIOLOGY OF THE SPINAL CORD.

The spinal cord or spinal marrow is that part of the cerebro-spinal system which is contained in the spinal canal of the backbone, and extends, in our domestic animals, from the head to a short distance behind the loins. It is an irregularly cylindrical structure, divided into two lateral symmetrical halves by fissures ; a superior and an inferior median, of which the former is the deeper. It terminates posteriorly in a pointed extremity, which is continued by a mass of nervous trunks (Cauda Equinæ). Running along the supero-lateral aspect of each half of the cord, is a third, though much shallower groove (supero-lateral), corresponding to the points of origin of the superior roots of the nerves, and dividing the inferior from the superior column. In the middle of the inferior column is a line formed by the exit of the motor roots of the nerves, and which is sometimes made to signify a further division into inferior and lateral columns.

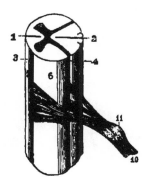

Fig 210.—Portion of the spinal cord showing its cut end. 1, Inferior longitudinal fissure; 2, superior longitudinal fissure; 3, inferior columns; 4, line giving origin to the sensory roots of the spinal nerves; 5, line of origin of the motor roots; 6, lateral columns; 7, superior columns; 8, motor roots; 9, sensory roots; 10, mixed nerve; 11, intervertebral ganglion on the sensory roots.

A transverse section of the cord reveals that it is composed of white matter externally, and of grey internally. The grey matter is arranged in the form of two crescent-shaped masses, placed back to back, and joined to each other by a transverse portion or *commissure* (grey commissure). The *horns*, of grey matter, extending upwards from the commissure, are longer but narrower than those which project downwards. The former extend to the surface of the cord, at the point where the sensory roots of the spinal nerves are connected with it ; the latter extend downward and outward towards the anterior roots, but stop short before reaching the surface. The inferior columns are still further connected by a portion of white matter extended across, between the grey commissure and the inferior median fissure. This is the *white commissure* of the cord. The cord is not of a uniform diameter throughout. Its dimensions vary according to the number and size of the nerves which originate from any particular point ; as, for example, in those parts which correspond to the lower end of the neck and the posterior part of the loins, it presents two considerable enlargements, from which originate the great nervous trunks of the anterior and posterior limbs. The

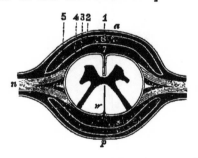

Fig. 211.—Transverse section of the spinal cord and its membranes. *a*, Inferior aspect; *p*, superior aspect; *n*, nerve; 1, dura mater; 2, 3, outer and inner layers of arachnoid; 4, pia mater; 5, ligamentum denticulatum; 6, arachnoid cavity; 7, subarachnoid space.

weight of the cord in middle-sized animals is, according to Chauveau—in the horse, 75½ dr. (300 grammes) ; in the donkey, 38 dr. ; in the cow, 55 dr. ; in the sheep and goat, 770 gr. ; in the pig, 1078 gr. ; in the dog, 539 gr. ; in the cat, 123 gr. ; and in the rabbit, 77 gr.

The spinal cord does not fill up the whole spinal canal. The latter contains, besides, a large venous sinus, fatty matter, the membranes of the cord, and the cerebro-spinal fluid. The latter is a serous liquid, which exists between the cord and its serous covering, on both its superior and inferior aspects, and is so abundant that the cord seems to float loosely within it. It is usually more abundant in old animals. Bernard remarks that, owing to the collection of blood in the sinus during expiration, and to its withdrawal through the suction power of the chest during inspiration, this fluid is thrown up toward the brain in a regular succession of waves; and that a prolonged expiration will accordingly lead to more or less vascular compression of the brain.

The spinal nerves—42 or 43 in number in the horse— arise each by two roots, a superior or sensory, and an inferior or motor. The former, which is the larger, has, at the point of exit from the spinal canal, a small ganglion, in which, according to Leydig, the nerve-vesicles are *bipolar* (having two prolongations). One prolongation of each cell is continuous with an afferent fibre, whilst the other is prolonged into one passing to the cord. Outside the spinal canal, the roots meet to form mixed nerves, which contain both sensory and motor filaments.

A knowledge of the minute structure of the spinal cord, and the mode in which its various parts are connected with the nerves, is an essential preliminary to the proper understanding of the purpose it fulfils, in connection with the various movements of the body.

The antero-lateral columns of the spinal cord, according to Van der Kolk, consist of white fibres running in great part longitudinally and parallel to each other; though, to some extent, arranged in a transverse direction. The former class seem to pass directly to the brain, and probably convey volitions to the motor ganglionic cells in the anterior horn of grey matter. Of the transverse fibres, some are merely the nearest of the longitudinal turned inward to gain the ganglionic cells; others consist of fibres passing across the inferior or white commissure, which is entirely made up of such decussating fibres; a third class are the fibres which form the roots of the motor nerves passing outward, usually in two bundles, toward the surface of the cord.

The anterior horns of grey matter contain superficially a number of marginal or circumferential fibres intermixed with small ganglionic cells. These fibres, which are derived from those of the column, are connected with the cells, and these in turn with larger ganglionic cells placed in the middle and anterior part of the horn. The large cells are arranged in groups, and each group, according to Van der Kolk, represents and presides over a muscle or a group of muscles, which act always in concert. These cells are *multipolar*, the prolongations going some to connect them with superficial small cells, or directly through the longitudinal fibres of the anterior column with the brain; some to the adjacent cells of the same group; some to join the cells in the posterior horn of grey matter; and others to form the roots of the motor nerves.

The posterior columns are, like the anterior, made up of (1.) longitudinal fibres externally, which become oblique internally before they pass into the grey, and (2.) of transverse fibres, which, passing outward from the grey matter, form plexuses amongst the longitudinal. Some of the longitudinal fibres are continuous with fibres of the superior roots of the nerves, which accordingly proceed directly to the brain without entering the grey matter: these are the sensitive fibres. The transverse fibres seem, on the other hand, to be the reflex roots. These latter pass through the substance of the superior grey horn to join some groups of

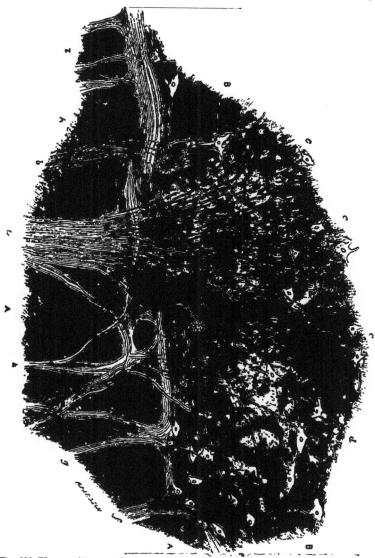

Fig. 212 (VAN DER KOLK) —Transverse section through the anterior horn and entrance of a nerve root. A A, Medullary matter; B B, grey matter of the anterior horn with its cells and nerve fibres; *a b*, two motor nerve-roots; *c c c*, multipolar cells into which the fibres of the nerve-roots pass; *d e*, another group of cells connected partly with the preceding, partly with one another, and which receive numerous filaments from the marginal fibres surrounding the grey matter; *f g h i*, grey radii; their fibres either pass at once into cells, as at *f*, or penetrate to the remote cells, as at *h*, or they form marginal fibres, as at *i*, all of which seem to be lost in remote cells, and decussate with the nerve root *a*, without passing into it.

ganglionic cells in their interior, and, through the prolongations of these, appear to be more or less directly connected
with the groups in the anterior horns.

The superior grey horns are made up in great part of
fine longitudinal fibres, and, as they are five or six times
thicker in the cervical and lumbar bulbs than elsewhere, it
is inferred that the fibres do not extend to the brain, but
become connected, at these points especially, with the
roots of the motor nerves. This is supported by the fact
that it is in the parts presided over by the above enlargements that reflex actions are most frequently and energeti-
cally effected.

The centre of the superior horns presents the nerve-cells
already referred to, and their surface has marginal fibres
with smaller cells similar to those of the anterior horns.
The grey commissure is formed of white fibres; these are
not, however, any more than those of the white commissure,
connected directly with any nerve-roots. They seem, in
part at least, to connect the ganglion cells which on the
different sides receive the reflex roots, and Van der Kolk
suggests that they may insure a co-ordinate movement of
the two sides of the body in reflex actions.

The spinal cord is possessed of all the functions predi-
cated of nervous centres, being itself but an aggregation of
such structures. This simple reservation may be made,
that it cannot directly affect or be acted on by the mind.
It may be stated shortly, that the spinal cord (1.) receives
impressions from all parts of the trunk, and generates motor
power as a result of such impression (reflex action); (2.)
transmits sensitive impressions to the brain; (3.) conducts
motor force from the brain to be distributed through the
corresponding nerves; (4.) harmonizes motor power, and
leads to a simultaneous and uniform action of muscles,

which contribute to the same movements in the body. Before noticing these, however, a few remarks will be given on the excitability of the cord itself.

It is held by many physiologists, that the columns of the cord correspond to the roots of the spinal nerves in this, that while the anterior columns are possessed of no sensibility the posterior are acutely sensitive throughout. M. Chauveau has carefully investigated this matter, having sacrificed as many as eighty horses in his experiments, and his conclusions are such as to modify greatly the above statement. The following are his results :—

1st, *On the Cord separated from its connection with the Brain.*—In this case, when the division of the cord was made in the upper or middle third of the neck, respiration was kept up artificially. On pricking or scratching the surface of the anterior or lateral columns, care being taken not to come in contact with the motor nerves, *no evidence of sensibility was obtained.* On treating similarly the surface of the posterior columns, *contractions* took place in the muscles supplied by the adjacent motor nerves. When slightly irritated, muscular contractions took place on the same side only ; when more actively irritated, similar effects were noticed on the opposite side, and on both of a more extensive character.

2d, *On the Spinal Cord connected with the Brain.*—Scratching of the anterior and lateral columns produced, as before, *no result.* Irritation of the posterior columns led to signs of pain and active (involuntary) muscular contraction. Varying the amount of irritation produced effects similar to those observed in the detached end, and the pain was most lively when the outer border of the column was touched, becoming less and less so toward the superior median fissure.

3d, *Excitability of the deep parts of the Spinal Cord.*—The cord having been carefully cut across in the interval between two spinal nerves, and the adjacent cut surfaces scratched, no effect is produced even in the case of the posterior columns, while the excitability of the outer surfaces of these is retained, though less vividly than before. Again, a needle may be passed in any direction through an unsevered cord without producing the slightest effect, provided always the surface of the posterior column is respected. If passed into the latter, pain and convulsions are noticed as it penetrates the outer layer, but after this it passes through the cord in any direction without effect. If, in the course of this experiment, the roots of the nerves were touched, or the nervous matter in their immediate vicinity, effects were induced exactly like those resulting from the direct irritation of those nerves.

4th, *After Death.*—If the surface of the superior column, or the root of the sensory nerve, be scratched immediately after the last beat of the heart, they do not respond, while, if the application is made to the motor nerve, active contractions may for some time take place.

Dr Brown-Séquard, who met with results identical with those of Chauveau, remarks, that while physiologists had erred in attributing a lively sensibility to the whole posterior column, at the same time the stimulus used in his and Chauveau's experiments might be insufficient to elicit the slight sensibility inherent in the internal parts of the cord. It may, however, be safely concluded that the sensibility of the spinal cord only exists to any extent on the surface of the posterior columns.

The reflex action of the spinal cord has been already referred to (page 387). It is the sole act of the cord as a nervous centre, is in itself entirely independent of the will, and consists in the reception of an impression through the sensory nerves, and the consequent generation of nervous force to be transmitted through the motor trunks. According to Van der Kolk, the transmission is effected through the ganglionic cells in the superior horn of grey matter, and the fibres which connect these with the motor cells in the inferior horn ; nervous force is generated in the latter, and conveyed along the motor nerves. Each group of cells in the anterior horns is understood to preside over a distinct muscle, or over several which act in unison, so that the application of a stimulus to a group leads to a natural and harmonious action. The groups being connected somewhat less intimately with each other, a more powerful impression leads to a simultaneous emission of force by two or more adjacent to each other, and to a more extensive though still natural movement. Moreover, the cells in the two lateral halves of the cord are connected in an analogous way, and accordingly reflex movements on the two sides simultaneously are by no means uncommon.

One of the most remarkable instances of this latter movement is in the case of the decapitated frog. He will remain quite still until an irritant is applied to some part of the body, when, if not severe, a foot may be mechanically raised to remove it ; if, on the other hand, an active irritation is made, the reflex movement will be much more extensive, and may even go the length of a leap. An instance almost quite as remarkable is that in which, from disease or injury of the spinal cord, its lower portion is cut off from all communication with the brain. No pinching or injury of the lower limbs can be felt, and no effort of the will can effect their movement ; yet, by tickling the sole of a foot, a sudden withdrawal of the limb may be still brought about.

The excitability of the cord, as exemplified in reflex actions, may be morbidly increased or diminished. In certain states of disease, as tetanus, or in poisoning by strychnia, it becomes so susceptible that the slightest sound or touch will often throw the whole voluntary muscular system into violent spasmodic action.

This reflex action is highly important as a protecting agent in case of irritants brought into contact with the body, but it is an agent in many other functions of a no less important character. The sphincter muscles of the bladder and anus are entirely dependent on a constant reflex action to keep them closed. Accordingly, when injury or disease has destroyed the functions of the lower part of the spinal cord, these relax, and the contents of the bowels and bladder escape involuntarily, and it may be without the knowledge of the patient.

The dilator muscle of the iris is under the control of that part of the cord which corresponds to the second dorsal nerve. Under the stimulation of the posterior columns or sensory roots at this part, the muscle contracts, and a marked dilatation of the pupil is effected. The transmission takes place through the sympathetic nerve.

By reflex actions the spinal cord seems to control, to some extent, the movements of the intestines, nutrition, and animal heat. Thus organic diseases of the cord are always accompanied by obstinate constipation, evidently dependent principally on the absence of the muscular movements of the intestines. The wasting of paralyzed parts may probably have a similar origin, since the circulation, and necessarily the nutrition, is greatly favoured by the muscular movements. Lastly, the heat of the body is largely dependent on the integrity of the cord, since, in broken-backed animals, the parts supplied by nerves given off below the lesion are usually cold, but always vary in temperature.

The spinal cord is a conductor of sensitive impressions to the brain, and it has been held that the transmission is effected through the posterior columns, while the anterior columns conduct motor power only. This, however, seems incorrect. The transverse section of one-half of the cord, and the posterior column of the other half, does not prevent the transmission of impressions of sensation to the brain. Bernard, who found such impressions conveyed even after cutting the anterior and lateral columns, rightly concludes that sensative impressions are largely conveyed through the gray substance; but Dr Nonat, Brown-Séquard, and others, have arrived at the conclusion, founded on numerous carefully conducted experiments, that not only do the grey matter and posterior columns convey such impressions, but that the anterior columns as well are to a limited extent possessed of a similar property. Dr Nonat, moreover, insists that not only do the anterior columns convey sensory, but that the posterior, in their turn, are capable of transmitting motor force.   In either case, however, this substitution takes place only to a limited extent.

The fibres of sensation cross each other in the whole length of the spinal cord, those of motion only in the medulla oblongata. Brown-Séquard cut transversely through one lateral half of the spinal cord, and then divided the motor roots of the lumbar nerves on the other side, that the opposite hind limb might not be contracted by reflex movement; and on pinching the limb on the same side as the cord was divided, the most marked symptoms of suffering were elicited. The opposite limb had at the same time, become insensible. The decussation of the fibres would seem to take place close to the roots of the nerves into which they enter, as the following experiments show. 1st, A part of the spinal cord was bisected longitudinally, so as to separate the two lateral halves of the cord from each other; the result was, that sensibility was destroyed in the parts supplied by nerves which leave the cord opposite the portion operated on. 2d, The right half of the cord was cut across in the dorso-lumbar, and the left half in the cervical region, with the effect of inducing a paralysis of sensibility, complete, or nearly so, in both hind limbs. The left half of the body and the left fore limb were in this case morbidly sensitive, while the right fore limb was almost destitute of this property. The crossing of the fibres conducting these impressions has been well illustrated by Lockhart Clarke.

Fig. 213 (LOCKHART CLARKE).—Diagram showing the crossing of the sensory fibres in the spinal cord. r, Right side; l, left side; s, section in loins; s', section in the neck; f, fibres imparting sensation to the right hind limb; f' to f''', fibres imparting sensation to the left hind limb.

We have remarked that the anterior columns are the chief conductors of motor power from the brain. Schiff, however, goes further, and ascribes to the anterior columns proper the transmission of such to the limbs only, while he particularises the lateral columns as performing a similar office with regard to the muscles of the trunk. They would seem also to be presided over by distinct nervous centres, so that they act in a great measure independently of each other. In support of this view, it may be mentioned that hemiplegia is not, as its name would indicate, paralysis of one-half of the body, but only of the muscles connected with the two limbs (fore and hind), while those of the trunk remain unaffected, as evinced in the normal continuance of respiration.

THE BRAIN.

The brain or encephalon is that part of the cerebro-spinal axis, which is contained within the cranium, and may be considered as simply an enlargement, consisting of numerous ganglia at the anterior extremity of the spinal cord. This part alone can hold direct intercourse with the mind; it only conveys sensations to the intellect, and gives rise to acts of will. Forming thus the highest part of the system, one rightly expects to find it attaining its greatest development in those creatures which are most exalted in the animal scale. This is true in general terms, and yet requires considerable qualification. It will be seen, by reference to the table given below, that if we judge of the intelligence only by the relative size of the brain, we must place the cat highest, followed by the dog, rabbit, ram, goat, and donkey. The horse will be degraded to a rank below all these, and placed on a level with the most stupid of ruminants. It would thus seem that nothing can be more liable to sources of fallacy than any attempt to predicate the extent of the intellectual endowments from the comparative development of the brain. Not to enter at present into the discussion of this subject, it may be stated that in no other animal does the brain, as compared with the body, attain a bulk at all proportionate to that of man—that in no other class is its development so great as in mammals; and that in no invertebrate animal is the brain so large as in the lowest specimens of vertebrata.

Table indicating the Weight of the Encephalon and Spinal Cord, as compared with that of the Body.—(Colin.)

Species of Animal	Weight of Body.	Weight of Brain.	Weight of Cerebellum.	Weight of Medulla oblongata, and Pons varolii.	Total Weight of the Encephalon.	Weight of the Spinal Cord.	Weight of the Encephalon and Spinal Cord.	Relation between the weights of the Encephalon and Body.	Relation between the weights of the Cerebro-Spinal Axis and Body.
Stallion, .	382·000	494	76	46	616	304	920	:: 1 : 620	:: 1 : 415
Gelding, .	380·000	559	77	39	675	300	975	:: 1 : 563	:: 1 : 389
Mare, .	408·000	510	71	34	615	269	684	:: 1 : 663	:: 1 : 461
Ass, .	175·000	316	45	24	385	159	544	:: 1 : 454	:: 1 : 321
Hinny, .	186·000	466	67	31	564	198	762 ·	:: 1 : 329	:: 1 : 244
Bull, .	293·000	403	52	33	488	177	665	:: 1 : 600	:: 1 : 441
Cow, .	332·000	416	44	30	490	225	715	:: 1 : 677	:: 1 : 464
Ram, .	46·000	112	15	10	137	52	189	:: 1 : 336	:: 1 : 243
Goat, .	37·500	95	15	12	125	48	173	:: 1 : 300	:: 1 : 217
Pig (fat), .	157·500	132	18	12	162	70	232	:: 1 : 972	:: 1 : 679
Sow (lean), .	74·000	85	11	9	105	44	149	:: 1 : 705	:: 1 : 497
Dog, . .	7·450	56	8	4	68	13	81	:: 1 : 110	:: 1 : 92
Cat, . .	2·342	20	4	2	26	7	33	:: 1 : 90	:: 1 : 71
Rabbit, .	2·135	...	...	...	8·5	4	12·5	:: 1 : 251	:: 1 : 171

That portion of the cerebro-spinal axis contained within the cranium has been subdivided into four portions, namely, the oblong medulla, which is continuous with the spinal cord ; the brain (cerebrum), which occupies the greater part of the cranial cavity, and is placed anteriorly ; the little brain (cerebellum), placed behind the brain proper and above the oblong medulla ; and, lastly, the *pons Varolii,* or *tuber annulare,* a ring-like eminence at the anterior part of the medulla, and occupying the point of union of the three above-mentioned parts.

The whole mass consists of a series of ganglia and nerve-fibres, the former being sometimes connected individually with others in adjacent parts of the encephalon, so that a short outline of its anatomy will be needful.

The oblong medulla presents an external appearance somewhat similar to that of the spinal cord ; it is, however, larger, especially in a sense transversely from side to side, and more flattened from above downward. The increase of bulk depends on the presence of a considerable amount of grey matter in its substance, on new fibres originating from this and passing in a longitudinal direction, and on numerous transverse fibres passing across between the various grey nuclei or coming from parts adjacent to the medulla. The superior and inferior fissures, which are so noticeable in the spinal cord, have almost disappeared in the medulla.

On the lower aspect of the medulla, we observe on each side of the median line a pyramidal mass (anterior pyramid) of white matter, wider and more diverging from its fellow in front than behind. It is not continued back into the cord, but makes its appearance on the lower aspect of the medulla, where many of its fibres seem to arise from masses of grey matter, and terminates anteriorly in the pons Varolii. Its internal fibres, which, according to Stilling, represent the motor-fibres of the lateral column, decussate in the median line with similar fibres from the opposite side; and it is here, accordingly, that volitional motor force crosses from one side to the other. The rest of its fibres pass up to the brain on the same side. It consists entirely of white nervous matter. The *restiform bodies* are two rope-like bodies on the outer side of the anterior pyramids. They are continuous with the posterior, and part of the antero-lateral, columns of the cord, but are made up also in great part of fibres originating from grey masses in the medulla, and of others descending from the cerebellum, into which it seems to be continuous, forming its posterior peduncle. On the superior aspect are seen two pyramidal bodies (posterior pyramids), which originate in some fibrous bundles on each side of the superior median fissure, and on reaching the medulla diverge from each other, becoming applied against the inner side of the restiform bodies, with which they seem to coalesce. Arnold has, however, traced them up into the brain proper. The interval between the posterior pyramids is the floor of the fourth ventricle of the brain, at the anterior part of which is the opening of a minute canal, which is continued along the whole length of the spinal cord between its grey and white commissures.

The *pons Varolii* is a quadrilateral elevation on the lower aspect of the brain, and forming the anterior limit of the medulla and the posterior of the brain proper. It is broader in front than behind, and has well-defined anterior and posterior borders. It is composed of grey matter, and of longitudinal and transverse fibres ; a large number of the latter pass upward to form the middle peduncles of the little brain.

In the cerebrum or brain proper, the grey matter is placed on the surface and the white within, so that it differs in this respect from the parts already noticed. Viewed from above the cerebrum is seen to be composed of two symmetrical halves (hemispheres), divided by a deep median longitudinal fissure. The surface is not smooth, but thrown into a great number of convolutions with intervening fissures, penetrating more or less deeply into the white substance, and thereby largely increasing the superficial extent of the grey matter. The grey matter of these convolutions preside over mental acts ; and accordingly in animals, with a large intellectual development, a proportionate increase in the number and depth of those convolutions is met with. A transverse groove on the lateral aspect of each, between the anterior and middle third, is known as the *fissure of Sylvius*. In this fissure are some semi-detached convolutions, known as the *island of Reil*. These are the first convolutions to appear in the lowest animals possessed of a cerebrum.

On turning to the base of the brain, a number of objects come into view, the principle of which we will enumerate, commencing from behind. From the anterior border of the *pons Varolii* start out two considerable columns, the *pillars of the brain,* or *crura cerebri (d d.)* These diverge in passing forward towards the right and left hemispheres, and leave between them a space known as the *intrapeduncular space.* The cerebral pillars are crossed in a direction from without forward, and inward by two rounded white bundles (optic tracts), (2″ 2″), which, on reaching the median line suffer an interchange of fibres with each other, so as to form the *optic commissure* (2′.) The optic tracts form the anterior boundary of the intrapeduncular space. In the intrapeduncular space is the pituitary body (*e*), the use of which is unknown. Beneath this is a mass of grey matter, perforated by a canal which leads to the third ventricle (infundibulum), close beside the last is a small grey mass (tuber cinerium), behind which is a white body (corpus albicans), and still more posteriorly the *posterior perforated space,* through which many fibres from the medulla pass upward to gain the cerebral hemispheres.

In front of the optic tracts are the anterior perforated spaces which lie beneath two striated bodies (corpora striata) placed on the floor of the lateral ventricles. In front of each anterior perforated space is the *olfactory bulb,* a considerable prolongation of nervous matter, grey without, white within, and containing a cavity which communicates with that of the corresponding lateral ventricle.

Turning once more to the superior aspect of the brain, and separating the hemispheres, we meet at the bottom of the longitudinal fissure with a mass of white matter (corpus

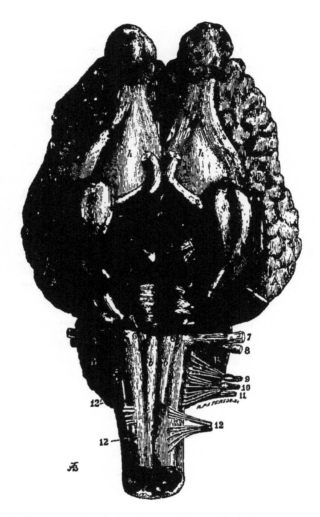

Fig. 214.—The lower part of the encephalon. A A A A, Hemispheres of the brain; B B, cerebellum; b, oblong medulla; D, pons Varolii; a, posterior cut end of the medulla; b a, anterior pyramids; c c, restiform bodies; d d, pillars of the cerebrum; e, pituitary body; f, corpus albicans; g g, mastoid lobes; h h, anterior perforated spaces; i i, external roots of the olfactory bulbs; 1 1, olfactory bulbs; 2 2, optic nerves; 2', optic commissure; 2'' 2'', optic tracts; 3 to 12, corresponding cranial nerves; 12', spinal roots of the spinal accessory.

callosum) consisting of fibres passing between the hemispheres. It is only about two-thirds the length of the hemispheres, and approaches nearer the anterior than the posterior extremity. Its function seems to be the connecting of the two hemispheres in intellectual acts.

If we cut off one of the hemispheres just above the level of the corpus callosum, and make a downward incision to one side of the latter, we open into the lateral ventricle of that side. The principal objects to be seen on its floor are :—1st, Toward the outer and anterior part of the ventricle a pear-shaped body (corpus striatum, 4 4) formed

Fig. 215.—Transverse section of the cerebrum, showing the lateral ventricles. 1 1, Hippocampi majores; 2 2, teniæ semicirculares; 3 3, choroid plexus; 4 4, corpora striata; 5, portion of the fornix.

of grey matter externally and white internally. This is connected with the substance of the brain and the *optic thalamus*. The latter is a considerable eminence at the posterior and inner part of the ventricle, where it is covered by the *choroid plexus* (3 3), an intricate network of minute blood-vessels projecting into the ventricle. A bundle of white fibres passes between the corpora striata on the two sides, and two commissures, one white and one grey, between the optic thalami. The lateral ventricle has two prolongations, one into the olfactory bulb and the other downward outward and forward into the substance of the hemisphere. In the latter are noticed a large curved eminence, the *hippocampus major*, or *cornu ammonis*, so called from its resemblance to a ram's horn, and an elongated white ridge, the fimbriated body. The *fornix* is a triangular white structure placed between the two optic thalami, and forming a portion of the floor of each ventricle. It is prolonged anteriorly and posteriorly by four pillars which descend into the substance of the brain. Each anterior pillar turns down to join the corpus albicans, and is reflected backward to join the optic thalamus. The posterior pillars are connected with the corpus callosum, and are continued as the fimbriated bodies. From the anterior part of the lateral ventricle a small opening (foramen of Monro) leads into the third ventricle.

The third ventricle is a cleft or interval between the two optic thalami and beneath the fornix. It is crossed by the three commissures of the corpora striata and optic thalami, opens at its anterior part into the lateral ventricle through the foramen of Monro, and at its posterior extremity through the aqueduct of Sylvius into the fourth ventricle.

Behind the third ventricle, and in front of the cerebellum, are some important structures. Applied against the posterior part of the optic thalami, and connected with these and with the anterior pillars of the fornix by two bands of white matter, is a small reddish body, the *pineal gland*. Immediately behind this, and above the aqueduct of Sylvius, is a large body partially divided into four by two fissures which cross each other at right angles. These are the corpora quadrigemina. These are larger in the lower animals than in man; the two anterior are largest in solipeds and ruminants, and the two posterior in carnivora. On each side of the corpora quadrigemina are two grey bodies bent upon themselves (corpora geniculata), which, with the former and the optic thalami, give origin to the corresponding nerves of sight. From the posterior part of the corpora quadrigemina two considerable white processes pass upward and backward into the little brain, of which they constitute the *anterior peduncles*. Stretched across between these peduncles, and above the anterior part of the fourth ventricle, is a white expansion containing also grey striæ, and known as the valve of Vieussens. In this valve may be noticed the fibres of origin of the fourth cranial nerve.

The *little brain* or *cerebellum* lies above the medulla and behind the brain. It is of considerable size in our domestic animals, and consists of two lateral lobes joined in the median line by an elongated worm-like eminence (vermiform process). The whole surface is covered by rugæ, with intervening depressions, in which the external grey matter dips deeply into its substance. The great depth of these grey folds gives to the central white substance an arborescent appearance, from which it takes the name of Tree of Life (arbor vitæ). In the substance of each hemisphere is a grey nodule indented at its edges, the dentated body of the cerebellum. This body contains ramifying white layers internally.

The white substance of the cerebellum is connected to adjacent parts by six prolongations of its substance, arranged in pairs, and known as its *peduncles*. The *anterior peduncles* pass downward and forward to join the corpora quadrigemina. The *middle peduncles* pass directly downward to join the lateral parts of the pons Varolii. The *posterior peduncles* pass downward and backward to become continuous with the restiform bodies.

In passing from the spinal cord to the medulla oblongata, the grey matter is met with more abundantly, becoming mixed up with all parts save the anterior pyramids. The superior horns become enlarged, diffused, and extended in a more lateral direction. Special grey nuclei likewise appear near the superior aspect of the medulla, and give origin to the roots of the nerves, such as the glosso-pharyngeal, hypoglossal, vagus, auditory, and spinal accessory. The posterior of these are imbedded in the substance of the medulla, but more anteriorly they stand out as rounded grey masses on the floor of the fourth ventricle. Two pairs of the nuclei lodged in the restiform bodies demand especial mention. These are the *olivary bodies*. They consist externally of a white material containing a grey nucleus thrown into folds or indentations, and in this, in turn, is a central white matter thrown into diverging branches. These bodies, which are extremely rudimentary, are connected respectively with the roots of the facial and hypoglossal nerves.

The fibres of the posterior columns are generally held to enter the substance of the restiform bodies, and proceed directly to the cerebellum. Van der Kolk, however, affirms that they terminate in the medulla in those nuclei which form the seat of sensation, while the fibres of the posterior peduncles of the cerebellum which join the restiform bodies become transverse, and either pass directly to the same pillars on the other side, or terminate in grey nuclei within the medulla.

Of the lateral column some fibres join the restiform

bodies and ascend to the cerebellum, some decussate with their fellows in the median line and pass chiefly to the opposite anterior pyramids, while a certain number pass directly upward through the posterior perforated space to enter the substance of the hemispheres of the brain. Van der Kolk supposes that most of these fibres end in nuclei in the medulla which presides over respiration, and that the ascending fibres are conductors of impulses of volition from the brain to those centres.

The anterior columns seem to give up a portion of their fibres to the restiform bodies and cerebellum, while others decussate in the anterior pyramids like those of the lateral columns, and ascend through the pons to the corpora quadrigemina and cerebral hemispheres.

The *pons Varolii* is composed of numerous transverse and longitudinal fibres interspersed with much grey matter. The superficial white layer of transverse fibres is continuous with the middle peduncles of the cerebellum. Beneath this is a layer formed of similar transverse fibres, and the longitudinal fibres from the anterior pyramids, largely mixed up with grey matter. Still deeper is a layer of longitudinal fibres forming the continuation of the lateral columns.

The cerebellum may be looked on as a great ganglion, with the grey or ganglionic matter surrounding the white fibres on all sides except the lower, where the latter pass out as the peduncles. Of these peduncles the median passes chiefly into the two lateral lobes, the anterior into the posterior part of the vermiform process, and the posterior into its anterior and middle portion.

The principal masses of white fibres in the brain are arranged in three sets—the *ascending, transverse,* and *longitudinal.*

The *ascending* are those of the pillars of the brain, which, coming from the *anterior pyramids, lateral columns,* and *posterior pyramids,* acquire additional fibres in passing through the pons Varolii, optic thalami, and corpora striata, and from the latter diverge into all parts of the hemispheres. Fibres of a similar kind are derived from the anterior peduncles of the cerebellum, the corpora quadrigemina, and the corpora geniculata.

The *transverse fibres* pass between the two sides of the brain, and are collected in great part into the corpus callosum. The white bands which connect the corpora striata and the optic thalami belong to the same order.

The *longitudinal* fibres are represented in great part by the fornix, the fibres of which, passing from the optic thalami, join the corpus albicans, whence they turn upward and backward in the median line, and ultimately bend down to become connected on each side with the hippocampus major and fimbriated body. At its posterior part it is connected with the pineal body. A layer of longitudinal fibres on the upper aspect of the corpus callosum, and several other smaller bundles of the same kind, likewise belong to this class.

The oblong medulla performs functions similar to those of the spinal cord as a conductor of nervous force. These are more important, however, as it will be seen from its position that all truly sensory impressions from parts supplied by spinal nerves must traverse the medulla in their course toward the sensorium, and in all voluntary movements of such parts the motor impulse must pass through its substance to gain the nerves of innervation. The same may be said of all reflex motor power originating in the medulla itself.

As a nervous centre this portion of the brain likewise

performs functions in the highest degree important to animal existence. It presides especially over the two functions of deglutition and respiration, and accordingly its removal leads to a sudden cessation of breathing and almost instant death.

The component parts of this organ would seem, as regards sensibility, to have some relation to the corresponding parts of the spinal cord. Thus, Longet found that irritation of the anterior pyramids seemed painless, whilst the slightest touch of the restiform bodies led to acute suffering. In addition, the anterior pyramids are the sole conductors of motor power, in proof of which Magendie observed, after section of one of these, a complete absence of voluntary movement on one half of the body, the sensibility meanwhile remaining unimpaired. The fibres of the anterior pyramids, moreover, cross over from one side to the other in the median line—a fact explanatory of those cases of transverse paralysis in which, from injury of one side of the pons or ganglia of the brain, paralysis of voluntary movement ensues on the opposite side of the body. Motor nervous force, accordingly, from the two sides of the brain, decussates in the oblong medulla, as we have already seen that sensory impressions do in the whole course of the spinal cord. The crucial transmission of impressions to and from the brain is best illustrated in some cases of disease. Several cases are on record of paralysis of one side of the body co-existing with atrophy of the opposite side of the brain (Van der Kolk); and every one conversant with sheep is aware of the imperfect control over one half of the body, when the opposite half of the brain contains a large hydatid.

The vital importance of the oblong medulla depends chiefly on the functions it performs as the centre of the

respiratory process.   To effect an inspiration, a very complex series of actions are brought about ; thus, the muscles
of the nostrils and cheeks must be acted on by the facial
nerve, the larynx opened by the pneumogastric, the diaphragm contracted by the phrenic, which leaves the spinal
cord at the fifth aud sixth segments of the backbone, and
the muscles of the trunk thrown into action by the spinal
nerves in the region of the back, before a satisfactory inspiration can be brought about.   This is not all, however.
The action of all the muscles implicated must be simultaneous, not only with each other, but especially so with
those of the opposite side, and a perfect harmony must
exist as to force and duration of impression.   According to
Van der Kolk, the various nuclei of these nerves are intimately connected with each other by nerve-fibres, and
those on the two sides are similarly connected through the
numerous transverse fibres that exist in the substance of
the medulla.   He holds, moreover, with Sir Charles Bell
and Schiff, that the lateral columns of the cord preside especially over the respiratory movements of the trunk, and
that as these terminate in nuclei close to the origin of the
pneumogastric nerve, the intimate connection of movement
between these and the others referred to is easily understood   Here, then, as in the spinal cord, are groups of
ganglionic cells, forming the origins of different nervous
trunks, and so intimately connected by intervening fibres,
that under a single impulse they lead to a harmonious action
of muscles situated in widely different parts of the body.

The impression which leads to the respiratory process
seems conveyed principally from the lungs through the
pneumogastric nerve, and results from the presence, in the
pulmonary capillaries, of blood charged with carbonic acid.
The extensive connections of the oblong medulla, how-

ever, leads to its being affected by many other causes, and a sudden full inspiration results, as every one knows, from the sudden contact of the surface of the body with cold water. Even the normal respiratory movement does not depend on the pneumogastric nerves alone, since, when both these nerves have been divided, a slow and irregular breathing is still kept up.

Some influence may be exerted by the spinal cord over the muscles of respiration, which supply the trunk, since Richardson and Brown-Séquard mention that they have noticed the respiratory movements in newly-born mammals after the oblong medulla had been entirely removed. There can be no doubt, however, that the action of other parts is almost, if not entirely, subsidiary. Thus, if the whole brain be removed, care being taken not to injure the medulla, life may continue for a considerable time, and respiration may continue uninterrupted. A complete division of the spinal cord at the lower end of the neck paralyzes the thoracic muscles, so that the movements of the thorax are thereafter continued only by the diaphragm. Again, section of the cord in the upper part of the neck paralyzes the latter, and breathing is at once arrested. Respiratory movements of the face may still take place, but none whatever in the body.

Flourens has attributed to a small nucleus of grey matter, lodged in the floor of the fourth ventricle, and close in front of the point where the posterior pyramids diverge from each other, the function of presiding over respiration, and to some extent over the movements of the heart. The nucleus in question he called the *vital point,* from the circumstance that when it was pricked or otherwise irritated, respiration was at once arrested, and frequently the action of the heart ceased as suddenly. Brown-Séquard has, however, actually excised this nucleus without respiration being at once arrested or interfered with, and finds, moreover, that it has no power to arrest or reduce the heart's action if the pneumogastric nerves have been previously divided. He concludes from this that the so-called vital point is not essential to life, and that the fatal interference with respiration and circulation, resulting from a wound of this nucleus, is entirely dependent on the irritation of adjacent parts.

It results that the oblong medulla is the centre of respiratory movement, and, accordingly, any serious injury of this organ is instantly fatal, while the remainder of the brain may be entirely removed, if done with proper precautions, without immediate cessation of vital processes.

The process of deglutition, like that of respiration, is dependent on the oblong medulla, as is proved by its persistence after the brain proper and the cerebellum have been removed, and by the inability to swallow after the medulla has been disorganised in experiments. This complex act is effected through the facial, hypoglossal, glossopharyngeal, pneumogastric, and probably the spinal accessory nerves acting on the jaws, cheeks, tongue, palate, pharynx and gullet. The whole is a reflex act, and results from the successive and regular action of the above nerves on the different parts engaged. The nerves on the two sides must also act in perfect unison, and all this seems dependent on the peculiar and intimate connection of the grey nuclei from which they originate.

The olivary bodies, in the opinion of Van der Kolk, preside over the articulation of sound, and their partial or complete destruction, in the human subject, leads to a corresponding loss of the power of speech. The above author instances a number of cases in which the patients could understand but could not articulate speech, and in which examination after death revealed that these organs were either altered by disease, or otherwise partially or completely absent. This theory would seem further corroborated by the fact, that though in the domestic quadrupeds two of these bodies exist on either side, they are so rudimentary that veterinary anatomists have frequently denied their existence. Both in man and in animals they are connected with the nuclei of origin of the facial and hypoglossal nerves.

It is curious, as shown by Bernard, that the pricking of the floor of the fourth ventricle, between the roots of the eighth and tenth cranial nerves, causes diabetes mellitus (secretion of sugar in the urine). If the pricking is made a little higher, the secretion of urine is diminished, and it becomes albuminous. If the puncture is made still more forward, near the pons, and close behind the root of the trifacial nerve, it leads to a great increase of the salivary secretion.

The pons Varolii is the medium of connection between the three other divisions of the brain, and, accordingly, contains fibres from each of these parts. The difficulty of experimenting upon this organ, from its position and its proximity to other parts of the most vital importance to life, and the numerous functions it possesses as a conducting agent, renders it extremely difficult or impossible to ascertain its precise properties. Longet found that the irritation of the inferior parts was not painful, but produced convulsions of the face, limbs, and other parts. Interference with the posterior part, on the other hand, produced lively suffering. Brown-Séquard infers, from the analysis of cases in which this organ was diseased, that its central part is the conductor of sensory impressions to parts higher up, while the inferior portion is the conductor of volitional force.

As a nervous centre, it seems to preside to some extent over some cranial nerves, such as the facial and abducens, which are connected with ganglia in its substance. It seems, besides, to be the first cerebral centre in which any power of sensation resides. This is proved by the experiments of Flourens and Longet, in which, after removal of the cerebral hemispheres, cerebellum, optic thalami, corpora striata, and corpora quadrigemina, so as to leave nothing but the pons and medulla, the animal still cried out on its tail being pinched, and raised its paw to its nose when that part was irritated by ammonia. The subject sought to lie in an easy posture, and if disturbed immediately resumed it. When left alone, the animal invariably remained quite passive and motionless. When the pons itself was removed, no further response was made on the application of an irritant, but the creature remained quite dead to surrounding things, the only vital movements retained being those of the respiratory and circulatory processes.

The functions of the cerebellum are by no means well understood. It has been held by Gall and Spurzheim, that it presides over the reproductive function, and varies in size according to the activity of the latter. Facts, however, rather oppose this theory. Many animals which are notoriously salacious, as kangaroos and monkeys, have comparatively small cerebelli, and the relative weights of this organ in entire and castrated horses point in the same direction. M. Leuret took the weight of the cerebellum absolutely, and as compared with that of the cerebrum, in ten stallions, twelve mares, and twenty-one geldings. The result of the absolute weights is given in the following table :—

	Average.	Highest.	Lowest.
Stallions, . . .	61	65	56
Mares, ...	61	66	58
Geldings, . . .	70	76	64

The greater weight of the cerebellum in geldings is the more remarkable, that the cerebrum of the stallion is on an average heavier than that of the gelding. This tends the more to remove the suspicion of fallacy in the comparative examination.

Flourens, Hertwig, Longet and others, have removed the entire cerebellum of birds piecemeal, and found that when the middle layers were reached, the movements became violent and irregular, while after the removal of the whole, all power of springing, flying, walking, or standing was lost. Sight and hearing, with sensation, volition, and memory, were still retained, and the bird would struggle violently to escape from any loud noise or threatened blow. It could no longer control and harmonize the action of its muscles in its attempts to fly; but fluttered and reeled as if intoxicated. From these experiments Flourens inferred that the cerebellum controlled and harmonized the various muscular movements, while Foville, with quite as much reason, attributes the symptoms to the loss of muscular sense.

Brown-Séquard calls both of these theories in question, and insists that the want of harmony between the movements depends not on the destruction of the cerebellum, but on the irritation to which the neighbouring parts are subjected. In support of this he adduces a case in which the cerebellum was deeply wounded in the human subject, notwithstanding which the man could still walk steadily, and even ascend a ladder alone.

He is further of opinion that the cerebellum controls to

some extent the circulation in the cerebrum and its nutrition, and that injuries of the former will, by modifying the latter, lead to various abnormal influences in different parts of the body. Among these he enumerates amaurosis, vomiting, headache, dilatation of the pupils, local or general convulsions, epilepsy, and hemiplegia, all of which occur also from certain conditions of the viscera, as in intestinal worms. In each case, he supposes the brain and oblong medulla as the immediate agent, acting under the influence of the cerebellum or intestines, as the case may be. Many symptoms, accordingly, noticed as resulting from injuries of the cerebellum, are not attributable to it as a separate nervous centre, but rather to the influence it exerts on those adjacent.

The different parts of the brain proper seem entirely destitute of sensibility, and may be cut away piecemeal without inducing the slightest indication of pain. The irritation of certain parts, however, gives rise to convulsive movements of various kinds, and the most remarkable is that of turning or rotation of the animal on its longitudinal axis. It is found that irritation or transverse section of one pillar of the cerebrum near the optic thalami causes the subject to roll over and over from the wounded to the sound side, and this movement may continue uninterruptedly for hours or days. Injuries to various other parts induce a similar rotation, or a turning from one side to the other, after the manner of a horse in a circus. Brown-Séquard has tabulated the results of these injuries, as follows :—

Parts producing Turning or Rolling after an Injury on the Right Side.

TURNING OR ROLLING BY THE RIGHT SIDE.	TURNING OR ROLLING BY THE LEFT SIDE.
1. Anterior part of the optic thalamus. (Schiff.)	1. Posterior part of the optic thalamus. (Schiff.)
2. Hind parts of the crus cerebri. (Schiff.)	2. Some parts of the crus cerebri near the optic thalamus. (Brown-Séquard.)
3. Tubercula quadrigemina. (Flourens.)	3. Anterior and superior parts of the pons Varolii.
4. Posterior part of the middle peduncle of the cerebellum. (Magendie.)	4. Anterior part of the middle peduncle of the cerebellum. (Lafargue.)
5. Place of insertion of the auditory and facial nerves (Brown-Séquard and Martin Magron.)	5. Place of insertion of the glossopharyngeal nerve. (Brown-Séquard.)
6. Neighbourhood of the insertion of the lower roots of the pneumogastric nerve. (Brown-Séquard.)	6. Spinal cord near the oblong medulla. (Brown-Séquard.)

Brown-Séquard remarks, that in some cases turning may result from vertigo, but attributes it, in the great majority, to spasms of the muscles on that side to which the animal turns. The fibres conveying the impression are, in his opinion, a special set, and quite distinct from those which usually act under the control of the will.

The corpora quadrigemina and corpora geniculata, which are connected with the roots of the optic nerves, seem to preside over the sense of sight, and destruction of these by disease or otherwise invariably leads to blindness. Wasting of the eyes may likewise induce diminution of these organs. In either case, the effect takes place on the opposite side— thus, disease on the right side of the corpora quadrigemina produces blindness of the left eye.

Injury to one side causes, as already noticed, a rotary movement of the body, probably from the induced blindness and vertigo.

The optic thalami, notwithstanding their name, do not preside over vision ; and they may be removed without, to any extent, interfering with this sense. The removal of one optic thalamus, however, leads to continued rotary movements ; the animal standing and turning perpetually to the injured side. If both are removed, the animal can still stand, and even use his limbs in walking, but no longer turns to one side. The turning thus seems dependent on a want of balance between the action of the two organs. The true physiological importance of these bodies is by no means well understood.

The corpora striata are placed between the crura cerebri and the hemispheres, and contain many fibres passing from the one to the other. They may consequently be supposed to exercise some influence on sensation and volition, but what their real functions are has not been satisfactorily made out. A more or less complete paralysis of the posterior extremities followed their mechanical injury in the experiments of Colin and others ; the effects, when one only was injured, showing themselves on the opposite side of the body. This is, however, not invariable ; and the statement of Magendie, that animals, after removal of these bodies, show an irresistible tendency to move rapidly forward, has been contradicted by more recent investigators.

The hemispheres of the brain are acknowledged, on all hands as implicated in the performance of the higher mental acts in the human subject, and in those acts of memory and intellect by which the animals immediately below man are characterised. The size of the hemispheres, accordingly, and the amount of the superficial grey matter,

as determined by its thickness and the depth of the convolutions, bear an approximation, though not constant
relation (see Table, page 410), to the development of the
intellect.   In the highest fishes they are still extremely
rudimentary, their size not exceeding that of the corpora
quadrigemina (optic lobes).   In reptiles, their size is relatively much greater; and in birds, they have so much
increased as almost to hide the corpora quadrigemina from
view.   Amongst quadrupeds, the dog presents himself as
at once possessed of considerable intelligence and a large
cerebral development; but above all, in the elephant do we
meet with the greatest mental endowments, and, relatively
to the size of his body, the · largest cerebral hemispheres.
It need scarcely be added that of all animals the human
subject presents the best example of a large cerebrum and
a correspondingly powerful intellect.

In further confirmation of this view, it may be stated
that in man congenital deficiency or atrophy of the hemispheres is usually accompanied by a corresponding deficiency of mental power.   Sudden and severe injury of the
hemispheres, as in apoplexy, may instantly deprive the
subject of all power of mind.   Gradually increasing pressure, as in the case of hydatids, in one of the hemispheres
produces a stupor which constantly augments with the
development of the cyst and the absorption of brain matter.
Lastly, the removal of the hemispheres leads to results of a
similar nature.   This can be effected in birds without any
immediate danger to life.   In the language of Dalton,
" the effect of this mutilation is simply to plunge the animal
into a state of profound stupor, in which he is almost
entirely inattentive to surrounding objects.   The . bird
remains sitting motionless upon his perch, or standing
upon the ground, with the eyes closed and the head sunk

between the shoulders. The plumage is smooth and glossy, but is uniformly expanded by a kind of erection of the feathers, so that the body appears somewhat puffed out and larger than natural. Occasionally the bird opens his eyes with a vacant stare; stretching his neck; perhaps shakes his bill once or twice, or smoothes down the feathers upon his shoulders, and then relapses into his former apathetic condition." Common and special sensibility are retained, but the animal is unable to associate the impressions with the ideas of their ordinary causes or sequelæ. Thus, the pinching of a foot causes the animal to move uneasily once or twice from side to side, but does not otherwise affect him. A pistol discharged behind his back leads him to open his eyes and turn his head half round, but does not seem to suggest any idea of danger or injury, and he immediately relapses into his condition of quietude. Longet found that by moving a lighted candle before the eyes of the bird, in a dark place, the eyes and head often follow it from side to side, and the eyes are sometimes fixed for several seconds on an object in the quiescent state.

The power of connecting ideas received from external objects seems entirely lost, and memory does not appear to retain an impression from one moment to another. But a tolerable control is exerted over the voluntary muscles, and if thrown into the air, the creature is still capable of flying.

The hemispheres, thus, do not seem essential to sensation, or even to some acts of volition, but are especially designed to receive and retain impressions, to associate ideas, to draw conclusions from them, and to preside over those movements which require a deliberate act of the mind.

The physiology of the remaining parts of the brain proper is by no means well understood. The corpus callosum has been supposed to effect a connection between the two hemispheres in the more complex mental actions. It should be remarked, however, that cases are known of absence or destruction of one hemisphere, in which all the more common acts of the mind seemed to be quite normally carried on.

The fornix and commissures of the optic thalami and corpora striata are supposed to connect those organs on the two sides between which they are placed, so that the action on the two sides may be uniform; but their true function has been by no means satisfactorily ascertained. The functions of the pineal and pituitary bodies are equally involved in mystery.

The nerves originating from the encephalon are 24 in number, and arranged in pairs, which are named first, second, third, &c., counting from before backward. They also receive special names, according to their functions, or the parts to which they are distributed. They may be classified according to their functions; thus, 1st, *Nerves of special sense*—olfactory, optic, auditory, part of the glosso-pharyngeal, and the lingual branch of the fifth. 2d, *Nerves of common sensation*—the greater part of the fifth, and a portion of the glosso-pharyngeal. 3d, *Nerves of motion*—third, fourth, smaller division of the fifth, sixth, seventh, and twelfth. 4th, *Mixed nerves*—pneumogastric and spinal accessory.

The first pair, or olfactory nerves, arise from the olfactory lobes, two prolongations of grey matter from the anterior part of the cerebrum, and of very large size, in the domestic animals. The nerves pass, as numerous filaments, through two perforated bony plates, to ramify in the mucous membrane in the depth of the nasal chambers. It is acknowledged on all hands that these nerves preside over the special sense of smell, and accordingly we find them largely

developed in those animals which possess this sense acutely.
M. Bernard has, however, thrown some doubt on the question
whether these are the only nerves presiding over this sense.
His objections are principally founded on the case of Marie
Lemens, in whose brain the autopsy showed an entire absence of olfactory lobes. Singularly enough, the testimony
of those most intimate with this woman went to show that
she disliked very much the odour of tobacco ; that she was
fond of flowers, and smelt them like other people ; that she
worked in a kitchen, and smelt and tasted the dishes in the
usual manner. And, lastly, that in her last illness she
constantly complained of the disagreeable odour of her
abundant perspirations. Standing alone as this case does,
and being, moreover, a congenital infirmity, too great importance ought not to be attached to it, and stronger
evidence must be adduced before it can be acknowledged
that the fifth, the only other nerve which sends branches to
the nose, assists in any degree in the appreciation of odours.

OPTIC NERVES.—The second pair of nerves consist of
white bands, which, from the corpora striata, corpora geniculata, and especially the corpora quadrigemina, pass
downward round the outer side of the crura cerebri, turn
forward and inward to decussate with each other in the
median line, and proceed through special bony canals to
their respective eyeballs. Their expansion in the eyeball
(retina) contains numerous ganglionic cells, and must be
considered as a true nerve centre. At the point of decussation, some fibres pass from the cord on the one side to
that proceeding to the opposite eye ; some pass into the
cord which proceeds to the eye on the same side ; some
pass into the opposite cord, in which they proceed backward to the brain ; and a fourth set, coming from the eye
on the one side, crosses over and reaches the opposite eye

without becoming connected with the brain. This nerve presides exclusively over the sense of vision, and by these crossings of the fibres, and the resulting connection of the centres implicated, the unity of vision is most satisfactorily accounted for. These nerves likewise give rise to the reflex action by which the pupils contract. They are utterly insensible to pain, and if irritated, give rise only to the impression of sparks or vivid flashes of light. If divided, vision is lost, and the iris ceases to contract through loss of its accustomed stimulus of light.

AUDITORY NERVES. — The nerve of hearing originates from a grey nucleus on the floor of the fourth ventricle, close to the median line. In passing outwards, its fibres are a good deal scattered, and in the intervals pass numerous longitudinal bundles, which, with the intimate connection between its nucleus and those adjacent, in the opinion of Van der Kolk, enables this nerve to act in a reflex manner upon these, and on the occurrence of a loud noise, to place the whole body as it were instinctively in a position of defence. The two nuclei are, moreover, closely united by transverse fibres, so that a sound falling on the two ears simultaneously produces a single, and not, as might have been expected, a double impression. The ramifications of these nerves in the internal ear present numerous ganglionic cells, so that this, like the retina, is truly a nervous centre. These nerves can convey sound only, and may be destroyed without producing pain. Inflammation existing in the internal ear or adjacent parts, gives rise to such perceptions as buzzing, ringing, rushing of water, &c. It is remarkable that mechanical destruction of the internal ear on one side leads to the phenomenon of the animal turning continuously to that side.

Motor Oculi.—The third pair, the motor nerve of the eye,

is connected with the grey nucleus in the crus cerebri, the corpora quadrigemina, and the valve of Vieussens. It is distributed to the muscle which raises the upper eyelid, and to all those of the eyeball except the superior oblique, which is supplied by the fourth, and the external straight muscle, which is moved by the sixth. In section or paralysis of this nerve, accordingly, the upper lid falls, and the eye squints outward from the constant and unopposed action of the external straight muscle. This nerve is connected with the sympathetic through a ganglion in the cavernous sinus, and with the fifth nerve at the ophthalmic ganglion. Through one or both of these means it is supposed to influence the iris, since the pupil generally dilates when it has been divided. (See " Sympathetic Nerve.") The intimate connection of the roots of this and the optic nerve enables the animal instantaneously and instinctively to adjust the axis of vision to luminous rays coming from any direction.

Pathetic Nerve.—The fourth nerve, which is the motor of the superior oblique muscle of the eye, arises from the valve of Vieussens close to the corpora quadrigemina, so that it is readily affected by impressions on the optic nerve. When cut or paralyzed, the opposing muscle draws the inner portion of the pupil upward, and from the light falling on points of the two retinæ which do not correspond, a peculiar form of double vision is induced.

Abducens.—The sixth cranial nerve goes exclusively to the external straight muscle of the eyeball, of which it is the motor. Its section or paralysis is followed by squinting of the eye inward, from the unopposed action of the internal straight muscle.

Trifacial, Trigeminal.—The fifth cranial nerve has two roots—a sensory and a motor—which arise from the lateral part of the pons, the latter slightly above the former. They are connected with grey matter within the medulla oblongata, and the sensory root may be traced backward as far as the origin of the hypoglossal nerve. A ganglion (Gasserian) exists on the larger root opposite the great foramen, at the base of the cranium, and immediately in front of this it breaks up into three branches—the ophthalmic, superior maxillary, and inferior maxillary—which are distributed to the parts about the eye, the upper and the lower jaw. The ophthalmic enters the orbit, and supplies the various parts therein contained, sending branches to the ophthalmic ganglion of the sympathetic, the conjunctiva, muscles of the eye, lachrymal gland and duct, the mucous membrane of the nose, and the skin upon the forehead.

The superior maxillary division passes forward beneath the orbit, and through a special canal in the superior maxillary bone. It is distributed to the maxillary sinus, the upper grinding teeth, and to the skin and muscles on the side of the face, nose, and upper lip.

The inferior maxillary contains a considerable portion of the fibres of the large root, and the whole of the small one. The fibres of the latter are motor, and this division accordingly contains both sensory and motor fibres unlike the other two, which are sensory only. This portion is distributed to the muscles, teeth, skin, &c. of the lower jaw, and presides over the movements of the muscles acting on the latter, namely, the temporalis, masseter, two pterygoid, and the mylo-hyoid. A branch to the anterior part of the tongue presides over taste and common sensation in this part.

The trifacial is the great nerve of common sensation to the face, as shown by the entire absence of this property in the skin, eye, and nostrils when it has been divided. The ophthalmic and superior maxillary divisions are exclusively sensory; but they nevertheless exercise considerable influence over the facial muscles by presiding over their muscular sense. They are further the natural excitants of many reflex actions in the facial muscles. That they are not motor is proved by Longet's experiment of stimulating them with galvanism, which produced excessive pain, but no convulsions. In the same way the lower maxillary division confers sensibility on the muscles, integuments, &c., of the lower jaw—with this addition, that it is the motor nerve of the muscles of mastication. When cut on one side, mastication is performed with difficulty, and entirely on the opposite side, and the animal falls off rapidly in condition. This division supplies, moreover, through its lingual branch, the special sense of taste to the anterior half of the tongue, so that this sense depends on two nerves, the glosso-pharyngeal and trifacial. On the integrity of this nerve depends indirectly the activity of the secretion in the lachrymal and salivary glands; an irritant touching the eye, or a sapid substance the tongue, producing an impression on the nervous centres, which through reflex action leads them to secrete.

It is very interesting to notice the influence which this nerve exerts over the nutrition of several structures—the seats of special senses. The division of the nerve, or of its ophthalmic branch, causes immediate contraction of the pupil, and in less than twenty-four hours acute inflammation of the conjunctiva and opacity of the cornea, followed by a purulent discharge; the iris becomes involved and loses all power of movement, and sight is lost, though the optic nerve and retina are unaffected. These results come on much more slowly if the section is made between the Gasserian ganglion and the brain, which seems to indicate that they depend in greater part, though not entirely, on the fibres of the sympathetic received at this point

Indeed, the ganglia of the sympathetic (ophthalmic, spheno-palatine, and otic) situated on the course of this nerve, at the points where branches are given off to the organs of the senses, seem intended to exert some special influence over the nutrition of those parts. It is well known that if we cut the division going to the nasal mucous membrane, which has the spheno-palatine ganglion connected to it, an inflammation in that membrane supervenes which renders smelling impossible. Furthermore, the nutrition of the whole side of the face is impaired if the principal trunk of the nerve is divided.

FACIAL NERVE.—The seventh nerve arises at the oblong medulla from grey matter, which is intimately connected with the roots of the trifacial and auditory nerves, and with the anterior olivary body. It leaves the cranium along with the nerve of hearing, from which it is quite distinct, and is distributed to all the muscles of the head except those of mastication. It is essentially the motor nerve of all the muscles of the face, with the exception just referred to. At its root it is exclusively motor, but in its course it acquires some sensitive fibres by its anastomoses with the trifacial, through the petrosal branches of the vidian, and probably through the chorda tympani. In passing from the cranium through the petrous temporal bone, it gives off filaments to the tympanum and muscles of the internal ear, and consequently performs important functions in adjusting these to suit the sonorous vibrations.

When this nerve is cut, the whole side of the face is paralysed; the eye remains partially open from loss of power in the orbicular muscle of the lids, and is constantly exposed to irritation from dust and other agents; the sense of hearing is impaired, if the section is made within the cranium; the external ear hangs pendulous; the nostril is no longer dilated at each inspiration, so that breathing takes place in great part by the opposite side, and if both are cut in the horse, which breathes through the nose only, suffocation ensues; the lips on the injured side hang downward, and cannot be used for the prehension of aliments; and, lastly, the aliments masticated collect in pellets inside the cheek, from the inability of the buccinator muscle to bring them between the teeth. The movements of the tongue also become more restricted, and the sense of taste is impaired or lost, not from the filament it supplies to the tongue (chorda tympani) being a nerve of taste, but rather from the want of a perfect adjustment of the organ, which seems necessary to the proper exercise of the sense.

Bernard has found that this nerve presides to a considerable extent over the secretion of saliva. He cut down on the side of the digastricus muscle, and raised the chorda tympani, separating it from the lingual nerve. He placed a tube in the duct of the submaxillary gland, and found that when the nerve was stimulated, saliva flowed abundantly, but ceased quickly on the removal of the stimulus. This phenomenon was repeated as often as the stimulus was applied and withdrawn. The sublingual gland likewise secreted actively when the chorda tympani was stimulated, and stopped with the cessation of the stimulus. The filament from the superior cervical ganglion of the sympathetic, which is connected with the submaxillary gland, has some control over the secretion, since its stimulation led to a greater viscidity of the saliva, and after a time to its complete cessation. Stimulation of the facial nerve within the cavity of the middle ear leads to a free secretion from the parotid gland, though the galvanising of the chorda tympani at its exit from the bone has no such effect.

Glosso-pharyngeal.—The ninth cranial nerve arises from

the side of the oblong medulla within which its roots are intimately connected with those adjacent—as the trifacial, facial, vagus, and hypoglossal. It leaves the cranium by the large foramen at its base, and at this point has a ganglionic enlargement. Below this it sends communicating filaments to the facial and spinal accessory. It is distributed to the mucous membrane at the base of the tongue, to the soft palate with its posterior pillars, to the middle ear, and the anterior constrictor muscle of the pharnyx.

Longet and Dr John Reid have shown that this nerve is essentially sensory, and though its stimulation leads to movements in the pharynx and upper part of the face, these, after section of the nerve, can only be induced by stimulating the cranial portion. Some motor power, which it seems to possess toward its periphery, depends on filaments derived from the spinal accessory.

This is, moreover, the nerve of taste, for the root of the tongue, the soft palate and its pillars—for those parts indeed where this sense is most acutely developed.

The glosso-pharyngeal is the principal medium for the conveyance from the pharynx of that impression which gives rise to the reflex act of deglutition. Swallowing does not depend on a reflex act conveyed through this nerve alone, since it may still be induced by a stimulus on the fauces after its section. The unnatural stimulation of this nerve gives rise not to deglutition, but to vomiting, in those animals in which this act can be performed. Tickling of the fauces with a feather is accordingly often employed to effect this object.

Pneumogastric.—This nerve arises by numerous roots from the lower border of the restiform bodies, and leaves the cranium through the same opening as the last. At this opening it has a ganglion, and just below this it communicates by anastomosing filaments with the facial, spinal accessory, hypoglossal, and the inferior branches of the 1st and 2d cervical nerves. It has been called the *vagus*, from its extensive distribution, which is made to the respiratory tracts from the larynx downward, to the pharynx, œsophagus and stomach, to the heart, to the liver, and other less important organs.

To the larynx, windpipe, pharynx, and that portion of the gullet which is in front of the heart (tracheal portion), it furnishes the pharyngeal, and superior and inferior laryngeal nerves; to the heart, several cardiac twigs; to the bronchial tubes, additional filaments; and to the thoracic portion of the gullet, minute twigs from its terminal branches, which in turn ramify in the stomach, liver, &c.

The functions of the pneumogastric are no less complex than is the distribution; it not only contains motor and sensory fibres in its composition, but likewise acts like the sympathetic upon the stomach and bowels.

The recent experiments of Chauveau have thrown much light on the functions of this nerve in deglutition, and we shall attempt, accordingly, to give a summary of the results at which he arrived.

In horses, cows, sheep, and dogs, branches from the pharyngeal and superior laryngeal bestow motor power on the soft palate, pharynx, and œsophagus, as far as the lower end of the trachea. Movements of deglutition take place in these parts when those nerves themselves, their branches on the gullet, or the pneumogastric above their origin, are stimulated, whilst no movement ensues when the galvanism is applied to the latter below the point where the former are given off. In man and in the rabbit, on the other hand, the motor power of the tracheal portion of the gullet is derived from the inferior laryngeal nerve alone. The inferior laryn-

geal or recurrent nerve, which is given off by the pneumogastric opposite the base of the heart, and ascends along
with the windpipe to the larynx, seems to have no function
connected with deglutition in the dog, but a very important
one in the case of the horse. In the latter animal, it conveys to the oblong medulla, the impressions made on the
mucous coat, and which lead, by a reflex action through the
pharyngeal and superior laryngeal nerves, to a vermicular
contraction of its muscular coat. Accordingly, it is found
that though no stimulation of this nerve can induce contractions in the tracheal part of the gullet, its section on
both sides, or that of the pneumogastric nerves above their
origin, leads to a paralysed or ataxic condition of this part ;
so that when a pellet of food passes no contraction ensues.
Deglutition can still be effected through the power of the
pharyngeal muscles and those in front of the neck. In
the dog, section of the recurrent nerves does not in the
slightest interfere with the contractions of the œsophagus,
thus showing that the sensitive reflex filaments are in this
case contained in the pharyngeal and superior laryngeal
nerves.

The thoracic portion of the gullet—that between the
heart and stomach—is supplied by twigs from the adjacent
trunks of the pneumogastric. If the latter, then, or the filaments derived from it, are galvanised in the dead animal, a
vermicular contraction of the gullet takes place. Again, if
in the living animal the vagi be cut in the neck, and respiration be meanwhile facilitated by performing tracheotomy,
the whole gullet is paralysed in the case of the horse ; and
though the animal may feed greedily, the aliments collect
along the whole length of the œsophagus, and can only be
passed on slightly when a new pellet is forced in by the
contractions of the pharnyx. In the dog, the thoracic por-

tion of the œsophagus is alone paralysed, and- deglutition can still be accomplished with tolerable ease.

The pharyngeal branch is the principal motor nerve of the pharynx and palate, and assists as well in giving motor power to the tracheal part of the gullet. The superior laryngeal, too, besides its action on the gullet, is the sensory nerve of the mucous membrane of the larynx, and it is owing to its keen sensibility, that the contact with the latter of a solid body, or irritant gas or liquid, gives rise to violent and uncontrollable fits of coughing. It is the motor nerve to a single laryngeal muscle, the crico-thyroid. The inferior laryngeal nerve, besides conveying to the brain the impression for the reflex action of the tracheal part of the œsophagus, and conferring on the latter its somewhat dull sensibility, has for its principal function the conveying of motor power to the muscles of the larynx. It is distributed to all the muscles that enlarge the glottal orifice; and, accordingly, when cut, the freedom of breathing is interfered with, and during rapid inspiration a sound of a more or less shrill character is heard. The animal in this case becomes a *roarer*, and the disease known as "roaring" frequently depends on a lesion of this nerve. If the nerves are injured on both sides, breathing becomes correspondingly more difficult. This nerve has the further function of modifying the voice, by varying the tension of the vocal cords.

The branches of the pneumogastric furnished to the lungs are the channels through which is conveyed that peculiar impression which leads to the continuance of the respiratory process. It is not, however, the sole agent for conveying this impression, since the non-aërated blood in all parts of the body transmit a similar influence through the various nerves, and respiration continues for a considerable time after the division of both pneumogastrics. When a section of these has been made, there are first symptoms of impending suffocation from the paralysis of the larynx, but these soon pass off, as the desire to respire is in a great measure abolished. The respirations take place slowly and at long intervals, often only three or four per minute, and death takes place in a space of time varying from twenty-four hours to six days. The lungs are throughout in a state of splenization, having a dark purple colour, and a leathery, resistent feeling, being incapable of crepitating, and infiltrated with blood. They sink in water. This seems dependent on the stasis of blood within their substance, and the collapse of the air-cells from deficient inspiration. Inflammatory spots, which sometimes appear, are referable to the irritation caused by foreign bodies which have passed unchecked through the paralysed larynx.

The action of this nerve on the heart is not well understood. After its section the heart's action is usually accelerated; but this probably depends entirely on the excited condition of the animal.

The pneumogastric presides over the movements of the stomach: when it is divided the organ is to a great extent paralysed; and, on the other hand, galvanising its trunk causes rhythmical movements of the stomach. Its section does not interfere with the sensation of hunger. The action of this nerve on the gastric secretion is disputed. Bernard found that its section led to an immediate stoppage of secretion, the previously turgid mucous membrane becoming pale and bloodless; and this condition persisted for days, and any food in the organ underwent decomposition. Others contest this opinion, and believe the cessation of secretion to depend on the shock to the system and the absence of muscular movement, which stimulated the mucous membrane, by bringing it in continual contact with the aliments. Longet, moreover, found that small quantities of food may be digested after section of these nerves.

Spinal Accessory.—The eleventh cranial nerve arises from the oblong medulla, just behind the pneumogastric, and from the lateral aspect of the spinal cord, as far backward as the sixth spinal nerves. It emerges from the cranium with the ninth and tenth, and divides into two branches, one of which joins the pneumogastric, and the other is distributed to the muscles of the neck and shoulder. It is essentially a motor nerve acting on the cervical muscles, and, through its pneumogastric branch, on those of the larynx. Its action on the larynx is to render tense the vocal cords, in order to the production of voice. The fibres of the branch rendered to the pneumogastric would seem to incorporate themselves with those of the recurrent nerve, and with it proceed to the larynx. Its control over the voice is proved by the fact, that section of the branch given to the pneumogastric produces instantaneous and complete aphonia, though the latter nerve be uninjured. Bernard has found that the fibres which preside over phonation belong to the anterior roots of the nerve, being all in front of the first cervical nerve. After division on both sides of the roots derived from the spinal cord, the voice remained perfect as before, whereas, when those in front of the first spinal nerve were cut, the voice became at first hoarse, and as soon as the connection with the oblong medulla was destroyed, became completely lost. The same result followed the division of the anterior roots, though those derived from the cord were left intact. In some cases the subjects were preserved for months, and even for years, during which the condition of aphonia continued. If injured, they opened their jaws as if to cry, but a somewhat hurried expiration was the only result. If the nerve on one side is cut the subject gets hoarse, though not aphonic.

It is remarkable in connection with the control exercised by this nerve over the muscles of the larynx, that when the former is paralysed or divided, the loss of motor power in the latter is associated with inability to close the glottis; whereas, after division of the pneumogastric or recurrent nerves, the absence of motor power in the larynx is associated with a closed condition of the glottis, which the animal cannot obviate. In the former case, respiration is free, but pellets of food fall through the glottis during deglutition, and irritate the air-passages. In the latter, food does not readily enter the air-passages during swallowing, but inspiration is rendered difficult, as is well exemplified in the case of the *roarer*. According to Bernard, therefore, the fibres of the spinal accessory convey to the larynx the influence necessary to the production of vocal sound, which is to some extent incompatible with respiration, while the proper fibres of the pneumogastric, rendered to the laryngeal muscles through the recurrent nerve, preside, on the other hand, over respiration. In this we meet with a substantiation of Sir Charles Bell's opinion, that when an organ receives nerves from different sources, it is not for the purpose of increasing the nervous force, but to convey to it nervous influences of different kinds.

It seems probable that the external branch of the spinal accessory, which is supplied to the levator humeri and trapezius muscles of the neck, has likewise a purpose to perform, for which the cervical motor nerves furnished to the same muscles are not fitted; and Bernard is of opinion that it is in movements incompatible with respiration, as severe straining and the like, that they are had into requisition.

Hypoglossal.—The twelfth cranial nerve arises from the posterior part of the oblong medulla, in a line with the lower roots of the spinal nerves. It arises from a ganglionic mass in the medulla, is intimately connected with the posterior olivary body, and through it with the nerves adjacent. It passes out of the cranium through a special opening at its posterior part, and is entirely distributed to the muscles of the tongue, of which it is the motor nerve. Irritation of its trunk causes convulsive twitching of the tongue, and its section completely paralyses it. It has no sensory fibres at its root, but, like the facial and spinal accessory, acquires some of these from adjacent nerves in its course. It has no connection with the sensibility of the mucous membrane, farther than its power of placing it in a position convenient for the reception of impressions.

CHAPTER XVIII.

DISEASES OF THE NERVOUS SYSTEM.

General remarks.—Rabies canina, or hydrophobia.—Causes.—Geographical distribution; symptoms.—Dumb and barking rabies.—Postmortem appearances.—Rabies in the horse, ox, sheep, pig, and cat.—Epilepsy; its symptoms.—Dr Brown-Séquard's researches on epilepsy.—Description of convulsions induced artificially.—Treatment of epilepsy.—Catalepsy in the dog; in a wolf; in the horse.—Chorea, or St Vitus's Dance.—Dr Todd's views.—Symptoms.—Treatment.—Tetanus.—Traumatic and idiopathic.—Symptoms.—Nature; treatment.—Diseases of the brain.—Vertigo.—Congestion.—Megrims.—Erroneous views commonly entertained on this subject.—Symptoms.—Prevention.—Encephalitis phrenitis, or inflammation of the brain.—Meningitis, or inflammation of the membranous coverings of the brain.—Difficulty of distinguishing inflammation of the brain from that of its meninges.—Causes and symptoms of encephalitis in cattle.—Encephalitis in the horse.—Treatment.—Apoplexy, or extravasation of blood.—Coma.—Immobility, or sleepy staggers.—Softening.—Induration.—Atrophy.—Hypertrophy.—Dropsy, or hydrocephalus.—Tumours.—Diseases of the spinal cord.—Paralysis.—Hemiplegia.—Paraplegia.—Congestion.—Inflammation.—Softening.—Dropsy.—Louping ill in sheep.—Trembling.—Thorter ill.—Cancer of the spinal cord.—Diseases of the nerves.—Neuritis.—Neuroma.

THERE are several functional and symptomatic diseases of the nervous system in animals which invest the whole subject I have now to consider with special interest, though, as Röll has correctly stated, the list of idiopathic diseases of either the central or peripheral portions of this system is a much shorter one than the list of primary affections implicating any of the other organs of the body. The history of nervous diseases in the lower animals has yet to be written, and I cannot hope to do more than furnish a sketch of the knowledge hitherto acquired on this very important subject.

It may be well to enter, in the first place, on the consideration of a very important group of diseases which implicate the nervous system generally, but which are not usually characterised by any marked organic changes. If the limits of this work permitted, much could be said on the subject of morbid mental conditions in animals. Are horses, oxen, dogs, and other animals ever mad? Are they liable to aberrations of intellect, and to morbid fancies or hallucinations? The ferocity suddenly manifested by animals which, early in life, have been very docile; the extreme irritability of some, and the apparently gross stupidity and listlessness of others, often indicate deviations from the normal state of either instinctive or reasoning faculties. There are singular instances noticed of animals acquiring peculiar morbid tastes, which can only be explained as due to nervous disease, and there can be no doubt that just as we find the greatest diversity in the amount of intelligence possessed by different individuals of the same species, so may we have perversions of instinct or mind similar to those which are manifested by the human idiot or lunatic. The disease commonly termed madness in the lower animals is not of the class included under the general term madness in man. It is altogether a specific contagious malady, which we may at once describe under the usually accepted names—

Rabies Canina, or Hydrophobia.

This disorder occurs in an idiopathic form, in animals of the canine species—viz., the domestic dog, wolf, and fox. It is communicable to all other warm-blooded animals by contagion, being characterised at all times by a train of symptoms which indicate great nervous derangement, though we cannot trace anatomically the changes which occur in the nervous system in this disease. The term Hydrophobia applies only to a symptom of this disease as it manifests itself in man; it signifies a dread of liquid, and, although mad dogs have been called hydrophobic, it is rare to see any symptom of such a condition in the lower animals.

Causes.—Little or nothing is known of the causes which induce rabies, primarily, in a dog or wolf. Our notions of the dog-days bear testimony to the popular belief that excessive heat and prolonged thirst send dogs mad; but scientific men have shown that the disease is as common in winter as in summer, and that the malady is unknown in many parts of the tropics, whereas it is most prevalent over the temperate European continent.

The geographical distribution of rabies would indicate that it is principally due to contagion; and if we only investigate the occurrence of the disease in the United Kingdom, we shall find that the malady prevails most where the chances of communication are greatest. It is occasionally prevalent in England, and especially in the midland and southern counties; it occurs most frequently in Ireland; and it is an extremely rare disease in Scotland. Not a little has been said regarding rabies in dogs in Scotland, but I have never seen a case during eight years, and Professor Christison assures me that for many years all his attempts at securing a rabid dog in Edinburgh, through the police and other likely channels, failed him. From the multitude of curs to be seen in any city in Ireland with bits of wood tied to their necks, under the impression that they can thus be prevented from biting, we might expect a very different amount of rabies to that witnessed in Britain.

A theory has been started to the effect that rabies in the dog commonly originates spontaneously from the restraint put on the animal's sexual desires. M. Leblanc, of Paris, believes this, and recently drew attention to the subject at the Academy of Medicine. On the authority of M. Sace, Leblanc stated, "that in a portion of the course of the Danube, the Christians on the one side of the river have chiefly male dogs, and there rabies is common, while on the Mussulman side, where, as in the East generally, dogs of both sexes are left at liberty, the disease is unknown." Leblanc went on to say, that "fancy and pet dogs, which are kept under most restraint in this respect, and are well fed, are those most liable to the disease. The great preponderance of the disease is shown from his own practice. Among 10,710 dogs entered upon his books, 30 per cent. of the number were bitches, or a little less than a third ; but of 159 of this number the subject of rabies, only 25 were females—a mean of 1 female to 14 males." There appears to me to be another explanation to the latter fact —viz., that it is usually dogs that manifest fighting propensities and who are most likely to be bitten, whereas bitches are often not bitten even by a rabid dog. How can Leblanc and his disciples explain the great immunity from rabies in our sporting kennels ? Good feeding and restraint in the sexual desires should surely take effect on many a foxhound and pointer during the year, but such a result is never witnessed. A general nondescript epizootic influence has been blamed for outbreaks of rabies, but the most effectual way to check the progress of the disease is to insist on all dogs being muzzled, and such a measure has been carried out with great effect in Berlin, Vienna, Milan, and other important cities. For some years a tax on dogs was imposed in Berlin with a view to diminish the number of roving and homeless animals, but the number of cases of hydrophobia did not diminish until 1854, when muzzling was ordered, and strictly executed upon all dogs not tied up. From the year 1845 to 1853 inclusive, 278 cases of rabies (nearly 28 per annum) were verified at the Berlin Veterinary School; while from 1854 to 1861 inclusive, only 9 cases have occurred, and none of these since 1856.

The virus of rabies is a fixed one, and discharged from the body in the saliva. It is most certainly introduced into the bodies of animals by the rabid creatures who, from natural cunning and ferocity, aim at the most vulnerable points of the human frame or other living thing. Wolves attack men on the face and neck, and the bite of a rabid wolf is thus more commonly followed by the development of the disease than the bite of a rabid dog, which is inflicted through clothes, &c. One great cause of the prevalence of rabies in various parts of the Continent, is the ready communication of the disease through wild animals. In Britain a rabid dog may attack a flock of sheep or fallow deer, but in the French and German forests he meets with victims capable of transmitting the virus *ad infinitum.* These are important considerations in relation to the causes of hydrophobia. According to Tardieu, 55 per cent. of the bites by rabid dogs prove effective in the transmission of the disease. This is very different from Hunter's statement, to the effect that only 5 per cent. of the persons bitten contract hydrophobia. Tardieu says that this is vastly under the truth, and only to be explained by including non-rabid bites.

Symptoms.—Two forms of rabies have been described,—the *dumb* and the *barking* rabies. They are only varieties of the same condition, dependent on peculiarities, so far as the symptoms are concerned, in particular cases.

There are four stages of the disease,—the incubative, the period of invasion, the critical stage, and the stage of decline and death.

The period of incubation varies in a most remarkable manner; and the result of an inquiry into 224 cases proves the appearance of the disease in less than a month in 40, from one to three months in 143, from three to six months in 30, and from six to twelve months in 11 cases. In children it has been as short as fifteen or even thirteen days, and rarely extends beyond twenty-five or thirty days.

The period of invasion is marked by an animal manifesting rather strange habits and desires. It may be more than ordinarily affectionate to its master, or it may begin by showing signs of discontentment. Restlessness, a capricious appetite, and indeed a desire to swallow filth and to lick urine, &c., are noticed. The animal's expression then changes, and is so characteristic, that when once seen it is never again forgotten. There is a peculiar, wild, anxious stare, which persons commonly attribute to some painful condition of the throat, or to a bone, &c. Very commonly ladies present themselves at the continental veterinary colleges, stating that their pet dog has swallowed a bone which has remained in the gullet, and when the professor sees the animal, he at once declares it rabid. In some cases there is dulness, laboured breathing, redness of the conjunctiva, dilatation of the pupil, salivation, and slight increase in the nasal secretion. Dogs which are seized with rabies, after having been bitten, manifest irritation in the region of the bite.

The third stage of rabies is indicated by the animal growing careless regarding the person and things amongst which he has lived. He takes to rove about, to howl piteously, and to bite whenever he can get a chance. The disposition to bite is increased at intervals, and there are paroxysms of apparently great suffering. Any object approached to the animal in this stage is bitten, and it is strange to see how steadily the eyes are kept on any thing moved within sight of a rabid dog. The dread of liquids, or hydrophobia, formerly declared as a symptom of canine rabies, does not exist, though there is difficulty in deglutition, whether of fluids or solids. Every kind of filth is however seized, and in part swallowed. The bark is very peculiar at this stage, and consists rather in a peculiar howl. There is redness and dryness of the visible mucous membranes and intolerance to light, shown by the eyes being kept partially closed.

The animal is then seen to writhe, to suffer from paralysis, its tail dropping between its hind legs, and the pulse is slow and irregular, breathing laboured, and convulsions occur. These fits are first partial and then complete, killing the animal about the fifth or sixth day from the first appearance of symptoms.

The form of dumb rabies is more rapid, and is characterised by paralysis of the lower jaw, discharge of saliva, swelling of the throat, catarrhal symptoms, very rapid emaciation, and early death.

Post-Mortem Appearances.—There are no specific lesions in the bodies of animals that have died of rabies. The principal manifestations are turgid state of the muscles and internal organs, congestion of the brain and spinal cord, determination of blood, and extravasation in the pharynx and stomach.   The tongue is often swollen and bitten, and the stomach contains filth of every description.

Rabies in the Horse.—The early symptoms consist in restlessness, dilatation of pupils, and in mares or stallions there is evidence of active sexual excitement.   As the disease develops itself, there are cramps, convulsions, and intolerance of light.   There is the paroxysmal tendency to bite, difficulty of deglutition, modified voice when the animal attempts to neigh, and partial paralysis of the hind extremities supervene.   The animal then suffers from convulsions, cannot rise, and dies about the sixth day.

In cattle and sheep, the symptoms are precisely similar to those in the horse.   There is great disposition to grinding of teeth and striking with the horns, though also a tendency to bite.

In pigs, the earliest symptoms consist in severe irritation of the bitten part, which the animal rubs and scratches. There is a peculiarly wild haggard look, rough voice, laboured breathing, and great disposition to bite.   The animals tear up the straw, and seize large mouthfuls of it. Paralysis and emaciation speedily supervene, and the animal dies about the third or the fourth day.

In cats, the early symptoms are to a great extent overlooked, from the shyness of these animals.   They, however, become very savage, and from their tendency to scratch and bite the exposed parts of the hand and body, they are more dangerous to meet with than mad dogs.   The disease ends in death, as in the pig, about the fourth day.

Rabies in man and animals is fatal in all cases.   Hot baths and a host of specific remedies have been tried, but invariably with bad success. Like other contagious diseases, it must be prevented by the segregation and slaughter of diseased animals. It is indispensable, when outbreaks of this malady occur, to place restrictions on the liberty of dogs, which should be tied and muzzled. No attempt should be made to treat rabid dogs, though it is also desirable to kill animals suspected of rabies, and which may have bitten human beings or other animals, without keeping them for a while to determine if they are rabid. In Scotland, dogs seized with ordinary apoplectic fits are commonly destroyed as rabid, and the ignorant prejudice still prevails that if a healthy dog bites a human being, and that animal afterwards happens to contract rabies, the human being will fall a victim to the disease.

EPILEPSY.

This is a disease which we have a great difficulty in defining. It occurs in all animals, but it is specially common amongst dogs, and particularly young ones. It is characterised by sudden fits; hence the name ἐπιληψία, a seizure.

Symptoms.—An animal in apparent health, or at all events quite calm and conscious the one moment, is seen to stagger, stare, and begin to champ with its jaws the next. The mouth foams, the muscles of the neck contract, the head is jerked upwards or drawn round to one side, the muscles of the trunk then contract spasmodically, and the animal falls, straining, struggling, and unconscious. Dogs are apt to cry out at first, but are quite dumb and listless during the seizures. The head is jerked violently, and the whole body severely convulsed. Fæces and urine are discharged involuntarily, and the animal breathes with difficulty. The mucous membranes are red and congested, and the heart

beats violently. The convulsive phenomena speedily subside and the animal regains its feet, and if loose is apt to
run from any one near it, or, in other cases, to fall into a
deep sleep. In severe forms of epilepsy, the convulsions
recur after, and the animal soon dies. In others, the fits
grow weaker and less frequent until they disappear altogether. Each fit may last from a few minutes to about
half an hour, though rarely as long as this.

Epilepsy is a disease due to a variety of lesions of the
nervous system, and to peculiar conditions of the blood.
Its attacks seem to commence often on the surface of the
body, or from some special centre within. In man some
part is first felt in a state of spasm, or is the seat of an un-
defined sensation varying in different cases. Such a sensation is termed the *aura epileptica*, and is of course not
usually recognised in the lower animals. It is evident,
however, from its manifestations, that epilepsy may be
either of centric or peripheric origin.

Dr Brown-Séquard published in 1857 some interesting
researches on epilepsy, recording specially the results of
experiments on animals. He says :—"I have found that
the following kinds of injury to the spinal cord are able to
produce epilepsy, or at least a disease resembling epilepsy,
in animals belonging to different species, but mostly upon
guinea-pigs :—

" 1st, A complete transversal section of a lateral half of
this organ.

" 2d, A transversal section of its two posterior columns,.
of its posterior cornua of gray matter, and of a part of the
lateral columns.

" 3d, A transversal section of either the posterior columns
or the lateral, or the anterior alone.

" 4th, A complete transversal section of the whole organ.

"5th, A simple puncture.

" Of all these injuries, the first, the second, and the fourth seem to have more power to produce epilepsy than the others. The first particularly, *i.e.*, the section of a lateral half of the spinal cord, seems to produce constantly this disease in animals that live longer than three or four weeks after the operation. After a section of either the lateral, the anterior, or the posterior columns alone, epilepsy rarely appears, and it seems that in the cases where it has been produced, there has been a deeper incision than usual, and that part of the gray matter has been attained. In other experiments, few in number, the section of the central gray matter (the white being hardly injured) has been followed by this convulsive disease. I have seen it but rarely after a simple puncture of the cord.

" It is particularly after injuries to the part of the spinal cord which extends from the seventh or eighth dorsal vertebra to the third lumbar, that epilepsy appears.

" Usually this affection begins during the third or fourth week after the injury. In some cases I have seen it beginning during the second week, and even one or two days before. At first the fit consists only in a spasm of the muscles of the face and neck, either on one or the two sides, according to the transversal extent of the injury. One eye or both are forcibly shut, the head is drawn towards one of the shoulders, and the mouth opened by the spasm of some of the muscles of the neck. This spasmodic attack quickly disappears.

" After a few days the fit is more complete, and all parts of the body, which are not paralyzed, have convulsions. According to the seat of the injury, the parts that have convulsions greatly vary. When the lesion is near the last dorsal vertebræ or the first lumbar, and consisting of a section of a lateral half of the spinal cord, convulsions take place everywhere, except only the posterior limb on the side of the injury. If the lesion consists of the section of the two posterior columns and a part of the lateral columns, and of the gray matter, convulsions take place everywhere without exception, but with much more violence in the anterior parts of the body. When the lesion exists at the level of the last dorsal vertebræ, and consists in a transversal section of the two anterior or of the two lateral columns, convulsions are ordinarily limited to the anterior parts of the body; but it is a very interesting fact that they are not always confined to these parts, the two posterior limbs having sometimes very strong tetanic spasms, at the same time that there are clonic convulsions in the anterior limbs. After a transversal section of the central grey matter, or of the whole spinal cord in the dorsal region, convulsions are limited to either the anterior or the posterior parts of the body.

"Convulsions may come either spontaneously, or after certain excitations. The most interesting fact concerning these fits is, that it is possible, and even very easy, to produce them by two modes of irritation. If we take two guinea-pigs, one not having been submitted to any injury of the spinal cord, and the other having had this organ injured, we find, in preventing them from breathing for two minutes, that convulsions come in both; but if we allow them to breathe again, the first one recovers almost at once, while the second continues to have violent convulsions for two or three minutes and sometimes more. There is another mode of giving fits to the animals which have had an injury to the spinal cord. Pinching of the skin in certain parts of the face and neck is always followed by a fit. If the injury to the spinal cord consists only in a transversal section of a lateral half, the side of the face and neck which, when irritated, may produce the fit, is on the side of the injury; *i.e.*, if the lesion is on the right side of the cord, it is the right side of the face and neck which are able to cause convulsions, and *vice versa*. If the two sides of the cord have been injured, the two sides of the face and neck have the faculty of producing fits when they are irritated. No other part of the body but a portion of the face and neck has this faculty. In the face, the parts of the skin animated by the ophthalmic nerve cannot cause the fits; and of the two other branches of the trigeminal nerve, only a few filaments have the property of producing convulsions. Among these filaments, the most powerful, in this respect, seem to be some of those of the suborbitary and of the auriculo-temporalis. A few filaments of the second, and perhaps of the third cervical nerves, have also this property of producing fits. In the face, the following parts may be irritated without inducing a fit:—the nostrils, the lips, the ears, and the skin of the forehead and that of the head. In the neck, there is the same negative result when an irritation is brought upon the parts in the neighbourhood of the median line, either in front or behind. On the contrary, a fit always follows an irritation of some violence when it is made in any part of a zone limited by the four following lines: one uniting the ear to the eye; a second from the eye to the middle of the length of the inferior maxillary bone; a third which unites the inferior extremity of the second line to the angle of the inferior jaw; and a fourth which forms half a circle, and goes from this angle to the ear, and the convexity of which approaches the shoulder."

Dr Brown-Séquard further on adds:—" The following description of these convulsions will show that, if they are not positively epileptic, they are at least epileptiform. When the attack begins, the head is drawn first, and sometimes violently, towards the shoulder, by the contraction of the muscles of the neck, on the side of the irritation; the mouth is drawn open by the contraction of the muscles of the neck, which are inserted upon the lower jaw, and the muscles of the face and eye (particularly the orbicularis) contract violently. All these contractions usually occur simultaneously. Frequently at the same time, or very nearly so, the animal suddenly cries with a peculiar hoarse voice, as if the passage of air were not free through the vocal chords, spasmodically contracted. Then the animal falls, sometimes on the irritated side, sometimes on the other, and then all the muscles of the trunk and limbs that are not paralyzed become the seat of convulsions, alternately clonic and tonic. The head is alternately drawn upon one or the other side. All the muscles of the neck, eyes, and tongue, contract alternately. In the limbs, when the convulsions are clonic, there are alternate contractions in the flexor and the extensor muscles. Respiration takes place irregularly, on account of the convulsions of the respiratory muscles. Almost always there is an expulsion of fæcal matters, and often of urine. Sometimes there is erection of the penis, and even ejaculation of semen."

Epilepsy may certainly be due to injuries of the nervous centres, or the nerves, and it may be due, and often is due, in young animals, to the whole nervous system suffering from defective nutrition, or the circulation through it of a poisoned blood, as in cases of fever, uræmia, jaundice, certain cases of poisoning, &c.

Treatment of Epilepsy.—This consists in treatment at the time of the fits, and general management of the animal with a view to the eradication of the disease. The epileptic seizures have been checked by arresting the transmission of impressions from the surface to the nervous centre. Dr Brown-Séquard alludes to this in the following terms :— " There is a great analogy between the *aura epileptica* in man, and the pain originating in the skin and face of my animals. In them, as well as in man (when there is a real aura), the trunks of the nerves seem not to possess the faculty of producing fits, whereas their ramifications in the skin, or in the muscles, have this power. In my animals, as well as in man, if there is an interruption of nervous transmission between the skin and the nervous centres, fits are no more seen, or at least their number is very much diminished. I have collected many cases of epilepsy with an evident aura epileptica, in which there has been either a diminution of the fits, or more frequently a complete cure, after the interruption of nervous transmission between the starting-point of the aura and the nervous centres. In these cases, the following various means have been employed with complete or partial success, either against the aura epileptica, or against its production :—1*st*, Ligature of a limb or of a finger; 2*d*, Sections of one or many nerves, and amputation of a limb, or of other parts of the body ; 3*d*, Elongation of muscles which are the seat of the aura ; 4*th*, Cauterization, by various means, of the part of the skin from which the aura originates."

I have found in dogs that the best means whereby to check the violence of a seizure, is to dash cold water on the head, or in severe cases, when the fits recur with violence, at short intervals chloroform may be given.

Schroeder van der Kolk regards epilepsy as usually due to, or associated with, congestion of the upper part of the spinal cord and of the medulla oblongata, and has ascertained that this condition is best overcome by belladonna.

or its alkaloid atropin, given internally over a period of
time. This has been attended with great success in
man.

Support by good food and the moderate use of stimulants, are of the greatest service; tonic remedies are used in
eradicating the disease, especially the oxide of zinc, oxide
of silver, nitrate of silver, sulphate of iron, &c.

CATALEPSY.

" A fit of catalepsy implies a sudden suspension of
thought, of sensibility, and of voluntary motion." Dr
Watson, who thus defines catalepsy, adds that the mental
faculties are in abeyance and the sensibility abolished, as
also the functions of voluntary motion; but the limbs are
not tied down by spasm; nor agitated by successive contraction and relaxation of their muscles; nor yet left, like
portions of dead matter, passively obedient to the laws of
gravity : they assume any posture in which they may
be placed, and that posture, however absurd, however (to
all appearance) inconvenient and fatiguing, they retain,
until some new force from without is applied to them, or
until the paroxysm is at an end. Catalepsy is by no
means a malady restricted to man. As far back as 1686,
Lochner described a case in a dog.* Leisering reported a
case a few years back which he saw in a wolf in the Berlin
Zoological Gardens. The limbs were held in any position
in which they were placed. Hering has reported a singular case in a horse, which suddenly stopped if at work, and
remained fixed with his fore legs propped out, and staring.
The seizures lasted for five or ten minutes. Cataleptic
symptoms are seen in a disease peculiar to horses, termed

* See Hering's Pathologie and Therapie, or Misc. Acad. Nat. Cur.

Immobility by the French authors, and which is described in the following pages.

CHOREA—ST VITUS'S DANCE.

A spasmodic twitching of certain muscles, occurring incessantly except when an animal is sleeping, constitutes chorea. Dr Todd has said that it is easier to say what chorea is not, than to describe what its essential nature is. "We may regard it as a disease dependent on a debilitated state of the system, which does not in any way arise from an inflammatory or hyperæmic state of any part of the great nervous centres or of other organs. Indeed, it is impossible to fix upon any particular organ of the body in which anything like structural lesion exists as a constant feature in cases of chorea. The disease is one of functional disturbance rather than of organic change, and this is borne out by the results of *post-mortem* examinations ; for almost without exception we fail to detect in those cases of chorea which terminate fatally any morbid alteration which, physiologically, could give rise to the phenomena ; and in the great majority of cases, we find all the viscera in a perfectly healthy condition—at least so far as we are enabled to make out with the means at present at our command. The structures which are obviously affected in chorea, are the nerves and muscles. Doubtless a morbid state of both exists ; but it seems most probable that the disturbed state of the muscles is excited and maintained by a deranged state of the nerves and nervous centres."

Puppies are more liable to chorea than other animals. It occurs in weakly dogs, and especially such as have been prostrated by a severe attack of distemper. In many cases it is restricted to a constant jerking of the lower jaw ; in others there is very marked twitching of the frontal and

upper cervical muscles ; in a third group of cases one of
the fore limbs is affected, and the jerking of one or both
fore extremities may be so severe as to interfere with the
animal's standing and walking, especially when it first
rises from the ground ; in a fourth and last group, most
of the voluntary muscles are implicated in the disease, so
that there is champing of the jaws, jerking back of the
head, spasmodic twitchings of the muscles of the back and
of the extremities, and the animal affected, unable to rest,
or to stand, or move at ease, is speedily exhausted by the
disease.

In addition to the twitchings of the muscles, we have
evidence in chorea of debility, defective nutrition, and a
tendency to wasting and prostration of the vital powers.

The disease has been seen in cattle and horses, but much
more rarely than in dogs, and is not often seen in other
animals.

Chorea is apt to become chronic, and in course of time
to disappear spontaneously.   In other cases it persists
with great severity, and much to the animal's inconvenience, if not suffering.

Treatment consists in the use of mild purgatives at the
commencement, followed by generous diet, and use of cold
water dashed daily on the animal's body.   Tonic medicines
are the most useful, such as the preparations of iron or of
silver.   The animals must be kept very quiet, but require
regular exercise.   Nux vomica, and strychnine have been
used with success, but they can be dispensed with in the
treatment of chorea.

TETANUS—TRISMUS—LOCKED-JAW.

The domestic animals, and particularly horses, are liable
to attacks of general, continued, and fatal spasm of the

muscles, both voluntary and involuntary, constituting the disease known as locked-jaw or tetanus.

There as two distinct forms of the disorder, differing in origin; the one succeeds the infliction of wounds and other injuries, and is termed *traumatic*; whereas the other is of primary origin or idiopathic. Traumatic tetanus follows surgical operations or accidental wounds which are often very slight. The spasms may supervene shortly after the occurrence of any injury, the time not exceeding fifteen or twenty minutes, though usually they are not observed until the wounds are nearly or quite healed. Predisposing causes, such as constitutional tendency, cold weather, low condi- tion, starvation, aid in the development of traumatic tetanus, though it is impossible to induce the disease at will by subjecting animals to influences which are at times sufficient to bring about an attack. Traumatic tetanus supervenes often after docking when the tail is cauterised, or after wounds of the feet or joints. It not unfrequently occurs in severe forms of cynanche or strangles after opening of the abscesses, or when suppuration does not occur freely. A cause of traumatic tetanus is the deposition of dirt in a wound, such as a particle of rust or a portion of steel from the breaking of an instrument. Punctured wounds are more likely to be followed by tetanus than incised ones, and we therefore have the disease after subcutaneous divi- sion of tendons or muscles, as in nicking. Traumatic tetanus not unfrequently follows after the occurrence of severe comminuted fractures.

Idiopathic tetanus is due to causes the nature of which is not always obvious. Cold is capable of inducing it, and to this cause I attribute cases such as I have seen of tetanus supervening suddenly when horses are clipped during very cold weather. It is seen in old, worn-out animals exposed to snow-storms and cold winds, and is also said to occur at any age from the influence of excessive heat. The condition of the alimentary canal materially influences the development of tetanus, and it has been known to occur after horses have eaten wheat or wheaten flour. Irritation of the digestive organs is sometimes capable of inducing the disease independently of other causes.

The tetanic spasms are termed tonic from their prolonged character and not being interrupted by relaxations, such as you have in clonic spasms or convulsive contractions. The term tetanus is used to indicate the spasmodic affection implicating the whole muscular apparatus. Sometimes the disease affects the muscles of mastication, and is then called Trismus (τριζω, strido). Special forms of tetanus have been described according to the symptoms induced by the violent contractions of special groups of muscles. The powerful superior cervical and dorsal muscles occasionally bend the spine backwards. This is called *opisthotonos*, and is occasionally met with in horses and cattle. An interesting case of this description was published in the first volume of the *Edinburgh Veterinary Review*. Mr Maclaren Kitching of Cupar, who reported the case, furnished the following particulars :—

" On July 4, 1854, I was requested to visit a black mare, the property of Mr Reid, late of Cruivie. The history I could glean was, that a day or two before, the ploughman was driving the mare in a cart, and when about two and a-half miles from home, she stood, broke out in a perspiration, and shook all over; he drove her home, and found her lame in the off fore leg. When seen by me, the mare could not well move the said leg forward; the pulse was regular; the muscles of the arm were of less size than on the near side; the triceps extensor brachii was flaccid, and not half the size of the near one. I considered it a case of lesion of the said muscles; gave a dose of physic, and ordered her to grass, as I considered that it was a case requiring time and rest. I mentioned that I would look at her in a few days, and see how she was doing in the park. Passing by, and not finding her in the park, I went to the farm, and saw the mare ; the pulse was perfectly regular ; respiration the same ; appetite good ; she seemed to have no difficulty in walking ; but when taken to the water-trough, she could not get down her head ; and upon more minute examination, I found her to be affected with tetanus without lock-jaw. I abstracted blood ; gave a smart dose of physic, as her bowels were costive, and applied a blister from the poll all over the spine. This was on the 9th July. Visited her on the 10th ; the medicine operating well ; blister discharging copiously ; gave more opening medicine combined with tinct. opii ; the head was drawn more upwards, and the mare was feeding well on grass, bruised corn, and bran mash. 11th and 12th.— Continued to give laxatives, combined with cocculus indicus, and changed alternately with small doses of belladonna. 13th.—Evidently worse ; her head much drawn back. I considered it to be a case of what is generally termed in the human subject, opisthotonos. To ease the muscles of the neck and back, the mare placed her fore feet in the manger, and rested her head upon the top of the rack. She never showed any symptoms of stiffness in her legs. On the 14th, applied a blister over her neck. 15th and 16th.—Her pulse increased in frequency, and the mare appeared to be getting worse. On the 18th, she lost the power of her legs. I called in the late Mr Dods of Kirkcaldy in consultation. He considered the treatment appropriate, and advised me to continue it with croton oil, so as to produce liquid evacuations, (she had always a capital appetite, even when down). Mr Dods recommended her to be placed in slings. I did so next day ; but she had lost the entire power of her fore legs ; and, when in the slings, she could not bear the least weight on them. She had the full power of the hind legs, and I had to let her down. Her head was so much drawn backwards, that when she was turned over to prevent the development of sores on the side she had been lying on for some time, the crown of the head and the posterior part of the quarters were the only parts that pressed upon the ground ; and the point of the withers was upwards of 18 inches from the ground. I continued the above treatment till the 25th, when she died. Twenty-four hours afterwards, I intended to make a *post-mortem* examination, but I could not get any one to skin her, or to assist me in any way ; so I cut off the fore legs, and examined the axillary plexus of each side, which was highly inflamed, and the nerves leading from them were in a state of inflammation even to their very centre."

When the body is bent forwards, as is often seen in poisoning by strychnine, the condition is called *emprosthotonos*. The term *pleurosthotonos* is applied to lateral tetanus in which the body is curved to one side.

The ordinary symptoms of tetanus are very characteristic. In the earliest stage an animal is seen to grind its teeth and champ with its jaws. There is often a considerable discharge of saliva, and in approaching or otherwise exciting the animal, the muscles of the face and neck are seen to twitch. The breathing is accelerated, nostrils expanded, nose protruded, and head elevated. The pulse is frequent, hard, and incompressible. On looking at the eyes they are found to be spasmodically drawn into the orbit, and the haw, or cartilago nictitans (commonly called "white of the eye"), is seen to protrude over the eyeball. The mouth may at first be opened rather freely, but as the disease advances, the strong masseter muscles prevent the jaw separating any distance. When the tetanic symptoms become general, the position of individual parts is regulated by the action of the more potent groups of muscles. Thus, the elevators of the tail being stronger than the pressors, cause an elevation. The anus is contracted powerfully. The limbs are forcibly extended, and cannot be flexed with freedom, so that the animal stands with obstinately outstretched limbs. The insertions of the levator humeri are distinctly traced, and the superior cervical muscles cause the appearance termed "ewe neck," and which once secured for tetanus the term "stag evil." Mr Percival says that voluntary muscles are those specially involved, but from the first we observe the partially voluntary and involuntary

muscles implicated.   Deglutition occurs with difficulty.
The peristaltic movement of the bowels is stopped, and
the urinary bladder firmly contracted.   The abdominal
muscles are rigid, and the intercostals do not move freely
in obedience to the will.

During the tetanic attack the animal's sufferings are
evidently intense, and there are periodic exacerbations of
great severity, brought on by exposing the animal to the
sun's rays, or to the annoyance of noises such as rustling
of straw, or to the inspection by people.   In the dark, and
when an animal is left in perfect quiet, the tetanic rigidity
is usually diminished, though convulsive twitchings occur
from time to time.

The disease may occur with great intensity; the expression of countenance denotes great irritation and pain, and
the animal sinks rapidly.   This is acute tetanus.   In other
cases the malady is of a less severe type, and if, by proper
care, the exacerbations are kept in check, it is not unfrequently seen that the spasms diminish in severity, and, if
the animal lives over the seventh day, it usually recovers.
During the progress of tetanus it is evident that the desire
for liquids is considerable, and the appetite not altogether
lost.   Attempts to eat are attended with aggravations of
symptoms, but large quantities of nutritious fluids, such as
linseed tea, milk, &c., are drunk if allowed to animals so that
they may drink when they choose.

Concerning the pathology of tetanus a great authority,
Dr Todd, says,—" We can only draw our conclusions
respecting the pathology of the disease from our knowledge
of the physiology of the parts concerned.   Now, reasoning
on this principle, it may be laid down that the phenomena
result from an exalted polarity of the centres supplying the
parts affected.   In the case of traumatic tetanus, the exalta-

tion of the polar state commences in the afferent nerves of the seat of the wound : if the tetanus arise from cold, the exalted polarity commences in the nerves of common sensation distributed to the exposed parts : from the periphery thus irritated the condition is propagated through the nerves to the centres, and the effects on the muscular system show to what portions of the nervous. centres the exaltation of the polar force is communicated. This, however, does not afford an adequate explanation of the production of tetanus; for peripheral nerves, and even nervous centres, are often subjected to great irritation without giving rise to tetanus ; and it is well known that it is impossible, even by severe mutilations, to produce tetanus in the lower animals : whereas a slight accidental injury (as when a horse picks up a nail) will often excite the disease in its worst form. It would seem that some peculiar state of the system, probably some peculiar condition of the blood, is a necessary precursor of this malady. Hence, no doubt, its greater frequency in warm and unhealthy. climates, in overcrowded and badly ventilated military hospitals, and among ill housed, ill clad, and ill fed infants. That tetanus may be produced through the blood is shown by the results of the administration of strychnine which irritate the tetanic symptoms in a very striking manner, so that you may at will develope the general phenomena of tetanus in an animal by giving him strychnine, or injecting it into his blood, but you cannot cause it by external injuries."

The *post-mortem* appearances of tetanus afford us no information of a positive character. Sometimes a particle of dirt is found near an inflamed nerve ; in other cases the brain and spinal cord are congested, but there is no constancy in the lesions met with, and we can furnish no facts of moment regarding them.

The treatment of tetanus has been spoken of as a " mortifying subject." The disease is by no means so constantly a fatal one in the lower animals as it is in man, but we are not prepared to attribute the recoveries to anything but the spontaneous efforts of nature. The greatest success has attended leaving animals perfectly quiet in dark loose boxes, with nutritious fluids to drink, and sufficient warm clothing to protect them from cold. Bleeding is objectionable ; purgatives of use, but cannot be repeated without unduly exciting the animal; narcotics are often decidedly prejudicial, and the one to be most avoided in the horse is opium.

Mr Horsburgh of Dalkeith, who has reported several cases of recovery after tetanus, says, " Like most of my neighbour practitioners, and for a number of years, I followed the school instructions and recommendations of our veterinary authorities, as published in their works, and, like most others, was equally successful—every case of traumatic tetanus died. I found the everlasting annoyance of balling, blistering, drenching, and clystering, aggravate the spasms to that degree that I had no doubt they hastened death. To save trouble in forcing up the head when the spasmodic action on the muscles of the neck rendered it nearly impossible, I tried, but with no better success, an opening into the œsophagus, by which I could pour liquid medicine into the stomach. I entirely changed my practice. Instead of constantly annoying the poor animal with medicines, I gave at first one large dose of physic—if the patient is seen early in the disease this is practicable— had him wrapt up from head to tail in four or five pairs of blankets wrung out of water about the heat of 200°, left him entirely quiet, locked up in a loose box, and his attendant to keep the key, allowing no person to go near or annoy him, unless in cases where any assistance was absolutely required. Thin sago or flour gruel given, and water, nearly boiling, to be poured along the spine outside the blankets every four hours. Bleeding largely at first, in cases of plethora, is, I think, always recommendable. This is the whole treatment I apply to these cases, and now find that I am generally successful."

Chloroform has been used in tetanus, and so long as animals are kept under its influence they are in a relieved condition; but the spasms come on with great intensity after its effects as an anæsthetic pass off.

Stramonium has been used with success by Buquiet, who steeped the plant in boiling water, and directed the steam, in large quantities, on the animal's body. Infusions of stramonium were injected into the mouth and rectum every hour. The treatment was commenced at seven o'clock in the evening, and in an hour the animal's skin became moist, and at nine o'clock there was considerable perspiration. The treatment was persisted in on the two following days, and the animal recovered.

I have tried cannabis indica, belladonna, hyoscyamus, &c., but with no decided result. The most opposite methods of treatment have proved successful, and a sufficient illustration of this is obtained from the cases of recovery after the shooting off a gun close to an animal's head.

DISEASES OF THE BRAIN.

CEREBRAL CONGESTION—MEGRIMS—VERTIGO.

When any organ becomes congested, the amount of blood in that organ is increased in quantity as well as stagnant in the vessels. It was once asserted that the quantity of blood in the brain could not vary in quantity, and this theory, enunciated first by Monro Secundus, and admitted by Abercrombie, was supported by experiments performed by Dr Kellie, who came to the conclusion,

1st, That in the brains of animals that have died of hæmorrhage, there is no lack of blood, but, on the contrary, very often a state of venous congestion.

2d, That congestion of the cerebral vessels is not met with in those cases in which we should most expect to find it; in persons, for example, who die strangled.

3d, That the quantity of blood in the cerebral vessels is not affected by gravitation ; in other words, that it remains the same, whatever may be the posture of the body and the position of the head.

The above conclusions have been demonstrated to be erroneous by Dr Burrows, who has shown that hæmorrhage has a most decided effect in depleting the cerebral blood-vessels, and in reducing the quantity of blood within, as well as upon the outside of the cranium. He has, more-over, proved that "the principle of the subsidence of fluids after death operates on the parts contained within the cranium, as well as upon those situated in the thorax or abdomen."*

Dr Watson says, we fall back "upon another principle whereby some of the difficulty and obscureness which attend certain affections of the brain and nerves may be explained—I mean the principle of varying pressure upon the nervous substance.

" Physiologists say that the cerebral matter is incompressible. This is another of the questionable assumptions implied in the foregoing theory. Upon what grounds the opinion may rest I am ignorant ; but whether the brain

* Lectures on the Principles and Practice of Physic by Dr Watson. Vol. i. p. 368.

is compressible or not, whether, that is, it be or be not reducible by pressure into a smaller compass, it is clearly capable of having different degrees of pressure applied to it, and of being pressed out of its ordinary form. We shall see hereafter, that by pressure exercised from within by the distension of what are called the ventricles of the brain, the convolutions on its surface are sometimes flattened, and the natural furrows between them nearly effaced. Pressure there certainly is, in what I shall have to describe to you as hypertrophy of the brain. There must be considerable pressure on the nervous pulp when blood is poured out within it from a ruptured artery in cerebral hæmorrhage. But the phenomena noticeable when a portion of the skull has been removed by the trephine, show very clearly that the encephalon sustains pressure from varying states of the circulation during perfect health. The surface of the brain, seen through the circular opening in the bone, is observed to pulsate, and to pulsate with a twofold motion. With every systole of the heart the surface protrudes a little, and it again subsides with the succeeding diastole. This shows that the tension of the arteries produced by every contraction of the ventricles of the heart, exerts a degree of pressure upon the contents of the cranium. But the brain has also an alternate movement corresponding with the movements of the thorax in breathing, rising with every act of expiration, and sinking with every act of inspiration. Now, during expiration, the blood escapes less freely from the head through the veins, and thus again vascular fulness is found connected with evidence of pressure on the parts within the head."

The lower animals suffer severely at different times from modifications in the amount of blood in the cranium, and greater or less pressure on the brain substance. In the horse there are a variety of forms of staggers or convulsive disorders due to functional or structural disturbance of the encephalon. In the chapter on Impaction of the Stomach in the Horse, at page 186 of the first volume of this work, I have drawn attention to the forms of staggers due to gastric derangement, and have described three kinds. One is characterised by delirium, a second by coma, and a third by interference with the voluntary muscles. The symptoms connected with these forms of stomach staggers are similar to convulsive attacks due to cerebral diseases of various kinds such as I have to notice in the following pages. I wish now to draw special attention to *Megrims*, the staggers of horses at work, and which has been mistaken for epilepsy.

Mr Percivall says, " By *Vertigo*, as synonymous with *Megrims*, I do not mean any simple or single symptom of giddiness which a staggered horse may evince ; but I mean an assemblage of vertiginous symptoms which suddenly attack, and as suddenly disappear after the manner of a fit, and to which horses all their lives may be at times subject, and yet never experience what we understand by staggers, *i.e.*, encephalitis, or phrenitis, or even coma. This makes me say megrims is a disease *sui generis*, though of what precise or definite nature, I am not at present prepared to give an opinion." In this paragraph which I quote from the Hippopathology, there are two errors. The one is attributing ordinary staggers to inflammation, and the other is regarding megrims or vertigo as a specific disease. That Mr Percivall, like all other authors who have written on megrims, has known nothing about the disease, is proved not only by his words to the effect that " *the pathology of megrims remains undeveloped*," but by his statement as to the causes of the disease. He says, " high or full condition, hot weather, exertion or agitation of any kind, may be said to be likely to produce a fit in a horse predisposed to megrims, although such causes are not in some cases recognizable. Harness-horses in particular appear subject to the disorder ; this may arise from the long continued constraint the bearing rein puts the head to. I knew a horse who had a fit of megrims every time he was put into harness, as if temper seemed to induce it." Any case of vertigo has been called megrims, though the staggering may only be a symptom of tumour in the brain or other organic disease.

I restrict the term megrims, as Mr Charles Hunting of South Hetton does, to a vertiginous affection only seen in animals at work and when driven with a collar. The vertigo never shows itself in the stable, or when the animal is ridden ; even in harness it does not occur, if an animal is worked with a breastplate and without collar. In Italy, a horse subject to megrims is said to have the "capo gatto," meaning really as mad as a cat, and it is well known that he is effectually cured by being used on the river side to draw along the rafts of wood floated down the streams to the sea, or even pulling boats up against the current. The fact is, these horses are never worked with collars, and signs of megrims are never seen in them. Some horses have such peculiarly shaped necks as to require a special kind of collar to prevent attacks of megrims, which are invariably due to pressure on the jugular veins. Heat, action, exertion, pulling heavy loads up steep inclines, are of course all causes calculated to aggravate the vertigo, but they cannot alone induce it.

The cases of staggers seen in saddle horses, or which occur at intervals even in the stable, are due invariably to organic lesions, and should not be confounded with that very simple and preventible series of symptoms observed in carriage-horses, from being unable to wear tight or badly fitting collars. True megrims occurs most readily in animals with obstructed jugular veins, from previous attacks of phlebitis, but such obstructions are rare now-a-days in this country.

The symptoms of vertigo or megrims come on suddenly, and often unexpectedly, so that accidents, especially in hilly districts, of a serious nature, often befal the persons driving horses thus attacked. The animal seized, elevates the head, has twitchings of the neck; sometimes the spasms attack the muscles on the one side, or all the superior cervical muscles are contracted, and owing to the peculiarly prominent eyeballs and wild look of the animal upwards, it has been called, when subject to these attacks, a "star gazer." The muscular twitchings of the face, turgid condition of the veins of the head, and dilated nostrils, are also marked, and with a sudden bound the animal falls against any obstruction or into any chasm without perceiving its danger. In fact the pupils are dilated and vision suspended during the next severe period of the paroxysm ; the reeling, delirium, or stupor ceases, and vision is suddenly restored on the circulation being re-established. Such attacks have been wrongly attributed to inflammation, phrenitis ; and grooms or farriers, not excepting even veterinary surgeons, are ready to jag the mouth in order to draw blood, or to bleed from any accessible artery or vein.

It will be usually found that an animal thus seized is being driven with a tight collar, or though the collar may appear deep enough when the animal holds its head down, it presses on the lower part of the jugular veins as the head is elevated in action, and especially in drawing a load up a hill. The way to relieve is therefore to push the collar forward, and if water be at hand, to pour some on the head.

Horses addicted to megrims have been driven with a

peculiar apparatus on the head to contain a damped sponge, but this is obviously useless, as the disease is entirely due to the malformation of the neck and the manner in which the collar rests on it.

ENCEPHALITIS—PHRENITIS, OR INFLAMMATION OF THE BRAIN —MENINGITIS, OR INFLAMMATION OF THE MEMBRANOUS COVERINGS OF THE BRAIN.

Although Mr Percivall and others have devoted distinct chapters to inflammation of the brain and of its coverings, we find in practice that it is impossible to distinguish the one from the other during the life of an animal. They occur principally in the horse and ox, and the morbid changes usually implicate one part of the cranial contents. The most common cause of either meningitis or phrenitis is injury such as concussion or fracture, and as an idiopathic affection occurring independently of injury, I regard phrenitis as almost an unknown disease. If animals are allowed stimulants in excess, however, we may expect congestions, extravasations of blood, and even inflammation of the brain ; such a cause rarely operates in inducing such lesions, if I except the singular cases first reported in the *Edinburgh Veterinary Review*, by Mr George Dundas, and which occur in cattle.

Mr Dundas refers to the malady as a form of chorea, but the *post-mortem* appearances satisfactorily indicate that the results of the cerebral irritation are congestions and inflammatory changes. The disease is due to the prevailing practice in different parts of Scotland, of giving "burnt ales" to cows in the neighbourhood of distilleries. The ale is given by steeping straw into it, and the animals will also drink it freely. They often sleep soundly after such a beverage, and sometimes symptoms of intoxication are manifest. The symptoms are as follows:—The head is turned singularly to the side, and is slightly elevated. The pupils are widely dilated, and the eyes have a remarkably wild appearance. On approaching the animals, they wink rapidly and tremble. There is marked heat of head, horns, and ears. When pressed with the finger in the axilla they fall instantly, and when pulled by the head they incline to turn over. The pulse is about 70 or 80 per minute. Mr Robert Morris has informed me that the symptoms of cerebral excitement are very great, and if the animals live on, chronic disease, due to exudations, &c., in the brain become confirmed. One cow, the case of which was reported by Mr Dundas, manifested symptoms of serious illness as the period of calving approached. Symptoms of delirium and interference with the muscular apparatus existed, and the animal had knocked off her horns in falling over in the stall and dashing about. After death all the organs are found healthy except the nervous centres, and both the brain and its membranes are found highly congested. This congestion often extends into the spinal canal, and the pia mater over both the brain and cord is the seat of red spots. The redness is either ramified, or is obviously due to blood extravasation. Clots of blood have been found in the lateral ventricles, and around the spinal marrow in the cervical regions. There is evidently softening of the brain substance as a direct result of this condition.

The violent symptoms usually attributed to phrenitis do not occur in the horse or other animal. In the early or congestive stage there may be delirium and paroxysms of excitement, or general convulsions. Symptoms of intense irritative fever supervene; the eyes are bloodshot; Schneiderian membrane red; mouth hot and dry; pupils contracted; breathing stertorous, and pulse wiry and frequent.

Any noise excites the animal, and the rustling of straw, or touching the surface of the body, indicates an increased sensitiveness to external impressions of various kinds. The symptoms, however, soon change, and indicate the aberration of the cerebral functions. We cannot accept Mr Percivall's description of the disease at this stage. He says : —" The frantic animal will rear both his forelegs into the manger, and in this posture stand, with his head erected, for several minutes perhaps, no person daring to approach the while, lest he should unexpectedly spring up or reel round, and fall upon the intruder. In a word, our patient is ' *mad*,' furiously so, in the worst sense of the word, as applied to staggers." Further on, Mr Percivall says :— " As the disease increases, instead of lying quiet as before, in a state of apparent insensibility after a throe, convulsions will follow so quick upon one another, that the patient will be kept in continual struggle, panting and perspiring, and perhaps foaming at the mouth, leading his attendants to believe he is not only phrenitic, but actually *rabid*." All this does not tally with the most reliable histories of cases of phrenitis especially due to injury, and the diagnosis of which could be relied on. Inflammation of any organ is, as a rule, associated with loss of function ; and accordingly, in well-marked encephalitis, after the paroxysmal or congestive stage, there is dulness, listlessness, and loss of consciousness. The animal stands with sunken and outstretched head, and either has difficulty in holding an erect position or lies. A certain degree of vivacity may recur, but it speedily disappears again, and there is decided loss of sensibility, and suspended function of all the organs of special sense. The pulse is apt to be full and throbbing at the submaxillary, and the breathing is loud, and stertorous as in coma. When the animal drinks it sinks its head in the pail, and tries to gulp down a little water.   It has no appetite, and the discharge of fæces and urine is very scanty. Prior to death the horse may be seized with convulsions, and knock himself about seriously, or he sinks exhausted and dies.

The duration of the disease varies from forty-eight hours to many days and even weeks.   In the chronic form there is partial paralysis, or peculiar modification in the action of certain groups of muscles.

TREATMENT.—The most valuable remedies in congestions and inflammations of the brain or its membranes are cathartics and cold applications to the head.   Clysters are also used, and saline diuretics administered.   Narcotics, such as opium, are contraindicated.   Blood-letting is more to be recommended ; and we find the opening of the temporal artery in favour with some veterinarians.   Mr Percivall says :—" When blood can be obtained from the temporal artery, that blood-vessel is to be preferred to the jugular vein.   In general, it is advisable to open both temporal arteries.   Should, however, even from both of them, the flow of blood be not free and abundant, the jugular vein must be had recourse to, it being absolutely necessary that blood in sufficient quantity should be extracted to produce symptoms of faintness, and it being highly advisable that this should be done as quickly as possible. Supposing the blood can be collected in a blood-can or water-pail—for this cannot on all occasions be accomplished —in general we shall find that from two to three gallons require to flow before this effect is produced, so much depending upon the size, condition, constitution of the horse, and other circumstances.   I used to consider the jugular vein to be quite as good a channel as, if not a better than, the temporal artery for blood-letting in affections of the head ;

but some striking cases I have had in my own practice have greatly altered my former opinions; and I find I am very much borne out in these altered views by the reports of others. At the same time, I wish it to be understood that arteriotomy is in no case to be confided in unless blood can be obtained from one or both temporal arteries in a full and fast stream. A dribbling or tardy current will avail nothing, and need not be persisted in."

My views with regard to blood-letting in encephalitis are that in the earlier stage it may be of use, but as the disease advances we must rely more on relieving the blood-vessels of the head by the direct application of cold, and perhaps using rubefacients to the extremities and trunk, so as to determine the blood from the head. Setons, blisters, &c., are only of use in chronic forms of the disease.

CEREBRAL APOPLEXY.

An animal, when seized with cerebral apoplexy, suffers from a sudden pressure on the brain, as the result of determination of blood and extravasation. The muscular apparatus is more or less completely paralyzed, but the heart and lungs continue to act. This condition of the system is termed the comatose. In reply to the question, What is coma? Dr Watson says :—" It is that condition in which the functions of animal life are suspended, with the exceptions of the mixed function of respiration; while the functions of organic life, and especially of the circulation, continue in action. There is neither thought, nor the power of voluntary motion, nor sensation, but the pulmonary branches of the par vagum continue to excite, through the medulla oblongata, the involuntary movements of the thorax. When this upper part of the cranio-spinal axis becomes involved in the disease, and its reflex power ceases, the breathing ceases also, and the patient is presently dead."

Under this head we must of course include many of the cases of parturition fever, or dropping after calving in cows, due to overfeeding and to a sudden accumulation of blood in the system on the birth of the young animal, which has largely drained the cow's system, and averted, up to the period of the expulsion from the uterus, a fatal accumulation of blood in the cerebro-spinal system. I have referred at length to this disease under the head Enzootic Disorders; and some of the cases in cattle due to drinking "burnt ale," referred to under Phrenitis, may be pure instances of cerebral apoplexy.

Our knowledge of cerebral apoplexy in the horse is limited. It is not a common disease; and by far the best case of the kind on record was published in the 3d volume of the "Edinburgh Veterinary Review," by Mr Parker of Birmingham, who says:—

" On Thursday, March 7, I was called in to see an aged bay gelding, the property of the Great Western Railway Company, and from the foreman I was able to glean the following history of the horse, up to this date.

" He was bought about five years ago, and was ill for a long time with influenza, but never seriously affected. He was put to work in due course, and beyond being rather bad-tempered, and, as the man said, 'curious in his ways,' there had been nothing to call for any remark till about the 1st of March this year, when he refused his food and became very stupid. This stupidity increased daily till the 6th, when he was sent to Birmingham by train from Leamington, at which station he had been working for two years. He now became paralysed, and was disinclined to move forwards, though he backed readily, and was conscious of

any order that was given him.   He walked from the railway
station to Hockley, where the stables are, a distance of 1½
miles, without assistance, though at any noise he ran back.
On seeing him, I found no very acute symptoms present.
The pulse was forty-four and natural in force, the eyes half-
closed and amaurotic, the sense of seeing *entirely* gone in the
near eye, and it was indistinct in the off eye.   He was feeding slowly, on my going into the loose box, and could turn
round without difficulty—though his movements gave me
the idea that the power of volition was but *partial*—both
fore and hind limbs were equally affected.   The bowels had
not acted for twelve hours, so I at once gave seven drachms
of aloes-Barb. and ordered clysters to be given every hour,
and some thin bran mash the only food to be offered to him.
I bled from both facial veins, and applied counter-irritants
to the poll and cranium.

" 8th. The horse is evidently better, he can see distinctly
with both eyes, has purged, and moves with greater freedom, showing now no inclination to run back.   He is rather
sick from the aloes.   The mash to be continued.

" 9th. Still improving, the dung pultaceous, his appetite
good, and symptoms of viciousness returning, as he begins to
look wicked, which I hailed as a return to his normal state.

" 11th. To-day I saw him walked out ; there were no signs
of paralysis, but he was very weak.   His appetite not so
good, and the sleepy look again apparent.   The dung was
getting hard too.   The poll and temples were stimulated,
and an alterative ball given.

" 12th. I was nearly kicked out of the stable this morning by my patient ; though weak, he tried to the best of his
ability.   His appetite was very good again, and he seemed
to be stronger.   He was led out for five minutes.

" 13th. No change from yesterday.

" 14th. To my surprise I found the animal lying down, quite unconscious, the eyes closed, mouth partly open, no sense of feeling anywhere present, and, in about eight hours from tumbling down, he died without a struggle.

" I thought that the brain had been affected sympathetically and sub-acutely at first—viz., on the 1st of March—and that probably there was a small clot of blood pressing on that organ ; and as he improved under treatment, and there were no violent symptoms, I hoped that the clot was becoming absorbed to some slight extent.

" On making a *post-mortem* examination, I found the brain in the state represented,—a large tumour pressing on the upper and posterior extremity of the left hemisphere. Whether the tumour was a partly organised clot I leave to abler hands than mine to determine, as, after showing it to Professor J. S. Gamgee, it was put into a box and sent off at once to Edinburgh. I may, however, venture to state my opinion, that the tumour was not of recent origin, as the brain was absorbed by the long-continued pressure of the mass. I have not disturbed the brain itself at all, so there may be some other pathological curiosity of which I am unaware at present."

" In my opinion, attacks of apoplexy, such as the one above described, occur in the horse most frequently as ·the result of injury, and the fact that the subject of the above case was vicious would appear to bear out such a supposition. The last case I had was in a mare, to which I was called in the last stage of an attack of colic, and by violently knocking herself about in the stall, effusion of blood had occurred below the crura cerebri, which gave rise to symptoms of coma and death.

" Mr Parker's case is one of old and, probably, relapsing apoplexy, and this opinion I form from the following appearances and microscopical characters. Lying on the posterior lobe of the left hemisphere, I found a solid and nearly spherical tumour, 2·2 inches long and 1·6 inches broad. On raising it, it proved completely detached from the cerebral mass, and divided into two portions, a lesser internal and posterior, and a larger external, the two applied by a flattened surface as if severed by a section.

The upper part of the tumour is covered by a dense white membrane, only partially adherent, and which appears to be a somewhat thickened portion of dura mater. Directly beneath and around the tumour are a number of flattened stratified scales. The separate layers are shiny and brittle, and prove by the microscope to be solidified lymph, without the slightest morphological character. The whole of the thickened portion of dura mater above-mentioned is lined by a dense coating of super-imposed strata of the same solidified fibrin.

A section through the middle of the tumour discloses that it is a solid mass of coagulated blood, undergoing the usual changes of apoplectic effusions. It is of a deep red colour, indicating its comparatively recent origin, though, from the colourless external strata, I should be led to suppose that the animal had suffered from apoplectic effusion at two distinct and distant periods.

The posterior lobe of the brain is evidently, to some extent, atrophied, though the amount of flattening is deceptive, as in all recent apoplexies, and sufficiently explains the irrecoverable cerebral disturbance.

On the whole, the brain appears to have been the seat of some determination of blood, but had a remarkably healthy appearance, with the exception of the two choroid plexuses, which are slightly thickened from a deposit of cholesterine, as commonly found in aged horses."

SLEEPY STAGGERS—IMMOBILITY—COMA.

Mr Percivall says, "The *coma* here intended to be introduced into veterinary nosology, is the *coma somnolentum* of human medicine, which, as near as a disease in man can represent one in a horse, is the sleepy staggers of old writers in farriery."

This malady is far more common abroad than in the United Kingdom, and has here been termed Immobility from the animal's indisposition to move, and it has been regarded as an acute hydrocephalus. Amongst the causes enumerated by authors are heat, obstructions to the jugular veins, mismanagement in feeding, &c. It will be found that breed has much to do with it, and it is rather a coarse kind of carriage-horse that is most liable to the disease. The animal's head is not usually a good one, and narrow across the forehead. The disease attacks middle-aged animals, and very rarely, indeed, either young or old horses.

Symptoms.—A general listlessness in the stable, disposition to be sluggish in harness, especially when kept standing a while without movement, a tendency to deep breathing and slow pulse, are amongst the earliest and most characteristic signs. The pulse is sometimes as low as 24 per minute, the respirations are also slow, the animal is in good condition, and seems to have a decided inclination to accumulate much fat. The organs of digestion are sluggish, and the discharge both of urine and fæces is scanty and rare. The slowness with which the animal feeds, the habit of seizing food between the lips, and then dropping the head in a sleepy fit, are very marked symptoms. When the disease presents itself in an aggravated form, you are perhaps told by the coachman that, whilst standing quietly in harness, on a smack of the whip to wake the horse up, he seemed to lose the power of his legs, and fell. When the disease has advanced thus far, we find in the stable that the animal heeds nothing except a loud sharp sound or the sudden admission of the sun's rays into a dark stable, and then it wakes from a state of drowsiness very suddenly, and may fall. Not unfrequently horses thus affected drop and break their knees. There is a peculiar fierce look and flaccid state of the facial muscles. The limbs are extended, and the animal obstinately stands ; indeed, the most curious positions may be given to the limbs, and they then remain as in cases of catalepsy. Thus one fore limb may be crossed over the other, and it is there kept ; a hind limb may be pushed far back or forwards, and the same indisposition to move it is seen. The hind limbs are apt to become the seat of twitching and even paralysis. The case assumes a chronic form, and it is rare to observe a fatal result except owing to some complication such as apoplexy, pulmonary congestion, &c. At times there are signs of delirium—paroxysm of spasm and great nervous excitement, but these pass off, and the animal is left dull and listless as before.

After death, accumulations of serum in the cranium, in the lateral ventricles, or in special cysts (Schöne, Tenneker), are occasionally met with. A state of anæmia, and sometimes of softening of portions of the brain, may exist, and tumour of the choroid plexus induce at times symptoms such as those I have just noticed. In acute cases of Immobility, there seems to be accumulation of liquid in the spinal canal as well as in the cranium, and it is in these cases that the hind legs are seriously affected.

TREATMENT.—This is often hopeless, but moderate and nutritious diet, purgatives, exercise, and rubefacients to the spine and limbs, are the remedies we may expect most direct benefit from. Nux vomica or strychnine and ferruginous preparations are of great value. Belladonna, in small doses daily, continued for some time, seems also to have a beneficial effect. From 3 to 5 grains of *nux vomica* have been given daily for some time to horses with this affection. Hertwig recommends croton oil as a drastic purgative in this disease. Bleeding is injurious. Viborg recommends the tincture of white hellebore. I have seen issues used with advantage, applied along the spine; and setons in the neck have been long in use, but are of questionable advantage.

SOFTENING OF THE BRAIN.

The French word for softening, *ramollissement*, has been applied to that condition of the brain in which its elements lose their firmness and cohesion. Softening is usually a result of inflammation, but it may be due to obstructions to the arteries, a species of arterial plugging or embolism which has been discovered often in the human cerebral vessels, and no doubt occurs sometimes in the lower animals. Softening of the brain substance may be connected with suppuration after an injury, or, as the result of an attack of phrenitis. Softening is usually a disease which attacks a part of the brain, and is therefore attended with symptoms indicating the loss of function of a particular part of the cerebral system. Loss of vision on one side, paralysis of the tongue, lips, and even of the muscles of the larynx, inducing severe roaring, may be its symptoms. The cases of softening due to arterial plugging and atheroma occur usually in old animals. Softening is an incurable disease.

INDURATION OF THE BRAIN.

Occasionally as the result of inflammation, and chiefly towards the surface of the brain, there is exudation and hardening. This condition has been rarely seen, and I have witnessed a form of induration which might be regarded as condensation of the nervous tissue under the influence of gradual pressure, owing to the slow development of a tumour. Induration is not diagnosed during life, and belongs to the incurable conditions.

ATROPHY OF THE BRAIN.

The brain may waste ; a portion of it may be absorbed, or the whole mass may diminish in volume. The cases of partial atrophy due to tumour, to the development of parasites, &c., are singular, from the fact that the disease advances far before any symptoms of the atrophy are noticed. The upper part of the cerebral hemisphere may waste away to a great extent, just as portions may be cut off without symptoms of disturbance, but anything like general atrophy is associated with loss of consciousness and paralysis. In man, atrophy of the cerebrum is accompanied by idiotcy, and in the lower animals by aberration of the instinctive faculties, and sometimes by viciousness. The general atrophy may be due to hydrocephalus.

HYPERTROPHY OF THE BRAIN.

We know nothing of hypertrophy of the brain in the lower animals, though the disease has been repeatedly described in the human subject. "When in these cases the skull is sawn through, the upper loose portion of bone starts up as if moved by a spring, and the edges of the bone remain widely apart." Dr Watson says that the hypertrophied and compressed brain is firmer and tougher than natural ; it contains but little red blood, and sections of it are seen to be unusually dry and pale.

HYDROCEPHALUS—DROPSY OF THE BRAIN.

The arachnoid, or serous membrane covering the brain, is not unfrequently the seat of an unnatural accumulation of clear watery liquid. This is not uncommonly seen in newly born animals, and especially in calves and puppies. Connected with this chronic cranial dropsy, there is malformation of the cranial bones. Many detached points of ossifications are seen in the broad membrane covering the encephalon ; and if this is pressed upon, the brain is affected and the animal injured, and perhaps destroyed. In the majority of instances the animals die. In adult animals hydrocephalus is rare, except as it occurs in sleepy staggers. Sheep, with sturdy, are said to die of hydrocephalus, or water in the head ; but it is well known that the cyst distended with fluid in this disease is of specific parasitic origin.

The cases of hydrocephalus in newly-born animals are usually, if not always fatal.

TUMOURS IN THE CRANIAL CAVITY.

A variety of products are developed in the cranial cavity with very different degrees of rapidity, and therefore inducing disorders of an acute or chronic character according to the conditions under which they are formed. The growths are either abnormal formations from the inner surface of the cranial bones or in the substance of the brain.

EXOSTOSES, OR BONY TUMOURS.

These are met with most frequently in cattle, and consist in hard, enamel-like products of the inner plate of the scalp, occasionally no larger than a bean, and some occasionally attaining the dimensions of a turkey's egg. These tumours are more or less globular, nodulated, with a distinct pedicle, and convoluted. The eminences or apparent convolutions are due to the manner in which the dura mater binds down the growth as it developes, and is fixed in depressions which are not unlike the furrows of the brain. The small tumours are found attached by pedicles, but the larger ones are usually floating in the cranial cavity, and pressing injuriously against the brain.

Some extraordinary cases have been recorded of growths which have attained the size of an ordinary ox's brain, and which have induced no apparent disorder until the animal's sudden death. One of these products—the largest on record —is in the Milan Museum, and was first described as an ossified brain. It has been the subject of many discussions, and it has been satisfactorily proved to be one of the ordinary bony growths from the inner plate of the cranial bones. The late Professor Alessandrini of Bologna carefully examined many specimens in the fresh condition, and all have been evidently connected at one period or other by a decided bony attachment with the skull-cap.

In the horse, the bony tumours consist in growths of a detinal-like substance, invading usually the temporal bones. They are probably aberrations in the development of dental pulps, just like the tooth products connected so frequently with fistula on the external surface of the temporal bones. These growths in the horse, like the bony cranial tumours of cattle, only induce symptoms when, either from their size or mobility, they press injuriously on the brain. They thus induce paralysis, blindness, and usually fatal disturbance of vital organs, such as checked respiration, &c.

DEPOSITS IN THE MEMBRANES OF THE BRAIN.

In connection with the membraneous coverings of the

brain there are occasionally deposits, especially in very young animals, and connected with a tubercular diathesis. This disease has been termed tubercular meningitis, and is by no means so well marked or frequent in the lower animals as in man.

Melanotic deposits and fibrous tumours occasionally occur, but are not usually of such a size in connection with the meninges as to lead to serious symptoms.

CEREBRAL TUMOURS.

In the brain the tumours which most commonly occur in horses are those of the choroid plexuses. These occur rather frequently in lowbred carriage-horses, and usually appear on both plexuses at once, and consist in abnormal accumulations of cholesterine, a non-saponifiable fat which occurs in rectangular scales. This cholesterine, with serum and amyloid or starch-like bodies, is found at first in the hypertrophied villi of the plexuses. In most old horses there is more or less enlargement of these vascular structures, but in some the growths acquire the characters of true cholesteatomatous tumours—that is to say, tumours consisting almost entirely of cholesterine, which is found packed in spherical masses, and these are surrounded by a somewhat dense capsule. The connective tissue in the substance of the tumours is scanty.

These tumours of the choroid plexuses grow slowly, and do not therefore induce severe symptoms until they have perhaps attained the size of a pigeon's egg. I have met with them as large as a hen's egg. They then usually give rise to staggering symptoms, which come on with great severity at intervals. When the fits are not on, the animal may appear quite healthy, fat, and in good working order; but under the influence of generous diet and work, violent exacerbations are brought on ; and whether on the road or in the stable, the affected animal becomes excited, raises its head, dashes forward blindly, and is seized with severe trembling. The head and ears are hot, pulse full and bounding, vision imperfect, and appetite suspended. With time the paroxysm is relieved, and the animal resumes its usual look of health. The disease is of course incurable, and is only aggravated by the methods of treatment commonly applied when the fits come on. Bleeding is especially injurious, and rather seems to favour a speedy fatal termination to the case.

HYDATID DISEASE.

Two cystic parasites are met with in the cranial cavity of the domestic quadrupeds. Cattle and sheep suffer from *cœnurus cerebralis*, a hydatid which develops in any part of the brain, and concerning which much has been said under the head Enzootic Disorders.

The second hydatid is the very common *echinococcus*, seen specially in oxen, and developed either in the meninges or in the substance of the brain. This hydatid, like the cœnurus, is due to the animals picking up with their food the ova of one of the tapeworms which infest the dogs' intestine.

DISEASES OF THE SPINE AND ITS COVERINGS.

PARALYSIS.

Loss of motor power and loss of sensibility may occur separately or together, and depend on functional or structural disorder of the nerves, or the centres with which they are connected. In brain disease we often have paralytic symptoms, but the spinal affections I have yet to describe are those in which the loss of power in muscles, or the insensibility of parts, constitute very special and pathognomonic symptoms. Paralysis may affect a part from injury to a single nerve, but more commonly we observe either the hind legs affected, as in paraplegia, or one side of the body, as in hemiplegia, or both sides of the body, whether partially or completely. We thus have the same loss of power varying in degree and extent; and that same absence of motion or sensation may indicate the existence of diseases varying greatly from each other. Paralysis, therefore, is but a symptom of disease, and the usual symptom in affections of the spinal cord and its meninges.

In all cases of disease or injury of the cord, or spinal canal implicating the cord, the paralysis occurs in the parts to which all the nerves are distributed, which originate or are connected with the cord behind the seat of injury or disease. If there be an injury to the sacrum, the tail alone may be paralysed; if the lumbar vertebræ are broken, the paralysis affects the hind quarters. If the cervical portion of the spinal cord is implicated low down, the fore limbs, as well as the hind ones, are deprived of power and sensibility; and if the cord is affected high up, near its point of union with the brain, the phrenic nerves can transmit no more impressions, the respiratory centre is injured, and the animal can no longer breathe.

The various kinds of paralysis differing principally in the extent to which the limbs and trunk are paralysed, are well illustrated by the symptoms which follow the various fractures of the spinal column.

When an animal falls on its head and breaks its neck close up to its attachment with the head, or when an animal is pithed, by dividing with a knife the medulla oblongata or upper part of the cord, there is instant suffocation and complete paralysis of the whole body. Movement and sensibility may be retained for a few seconds in the face, and the heart continues to beat under the influence of the ganglionic system of nerves, but all signs of life soon vanish.

Fracture and displacement of the lower bones of the neck may occur without paralysis, but usually both fore and hind extremities lose their power.

Fractures of the dorsal and lumbar vertebræ occur in horses from violent muscular effort during surgical operations. When the fracture affects the dorsal spines, the displacement is not so great as in fractures of the lumbar region, hence an animal may rise, walk to its stable, and only show signs of paralysis after many hours, and even days, have elapsed. Commonly with lesions in the lumbar region, there is at once irremediable paralysis of the hind quarters.

ACUTE RED SOFTENING OF THE CORD—MYELITIS—SPINITIS—ACUTE PARAPLEGIA.

This disease is somewhat rare, but a number of cases have been recorded in a more or less imperfect manner. During the year 1863 several cases came to my knowledge, but I observed one, in company with my colleagues, which can well serve as a basis for the history of the disorder.

Subject.—A five-year-old mare of medium size, nearly half-bred, and rather fat. Was accustomed to hard work in a heavy van. Had stood in the stable three days ; and on December 29, 1863, was taken out in the van for a mile or two. The day was cold, with a drizzly rain almost constantly falling. Nothing was noticed amiss with the animal until taken from the van, about six yards from the stable door. It then became excessively lame in the off hind limb, so that it was with difficulty it was got into the stable.

At 7.30 P.M. my father found it on three legs, with violent muscular tremblings, evidently in great agony, and shrinking from the slightest touch.

At 9 P.M. my father, Mr Law, and myself, visited the patient, and found paralysis of motion in both hind limbs, though sensation seemed to be retained. The animal could rise on its fore limbs, but only partially on the hind, these being bent, with the pasterns resting on the ground. The respiratory movements of both thoracic and abdominal muscles were normal. The hind legs were drawn up again when pulled back by the hand, and were thrown out slightly with the general movements in the frequent struggles of the animal. Pricking the tail with a pin caused its ready movement. The line of the lumbar and sacral spines was regular, and neither they nor the dorsal showed tenderness on being struck. The mare was in violent agony, with the pulse at seventy-two per minute, full, but soft, the skin drenched with perspiration, and the eye abnormally bright, as in colic. The expression of the face was haggard, and the head frequently turned to the flank, as in abdominal pain.

At 10 P.M. I found her in the same condition, excessively restless, and suffering severely. Gave an anodyne draught, which partially relieved the pain.

Examination per ano, detected healthy beating of the posterior aorta, and of both the iliacs on each side, and decided that no retention of urine existed.

By applying the ear over the heart near its base, I detected a rushing sound replacing the second sound of that organ. An hour later, Dr Arthur Gamgee accompanied Mr Law, and examined the heart carefully with the stethoscope.

The sounds were quite normal at the apex, but the second replaced by the beat already mentioned at the base. By pressing the instrument on the carotid, a double murmur was heard with each heart-beat.

The mare was now much easier, and the perspiration less free. A purgative was administered.

December 30th.—Patient much easier, pulse 45, bowels acting normally. Put a mustard poultice over the loins.

December 31st.—Mare much the same as yesterday, fed well, and the only noticeable symptom was the paraplegia. The owner now consented to have it destroyed.

January 2d, 1864.—At 11 A.M. Dr A. Gamgee and Mr Law examined the heart, which had been removed yesterday morning. Water poured into the aorta did not descend into the ventricle. Water poured into the pulmonary artery, flowed gradually into the right ventricle. The valves, notwithstanding, showed no symptom of structural change. The right ventricle seemed slightly dilated, probably from regurgitation of blood.

On laying open the spinal cord, its membranes appeared little altered from health. The cord in the dorsal and anterior half of the lumbar region appeared healthy. In the posterior part of the lumbar region the extremities of the anterior horns were much more vascular than is natural, while two or three inches of the extremity of the cord was quite softened and pulpy, and almost diffluent. Microscopic examination showed soft varicose nerve tubes, globules formed by their disintegration, and numerous exudation corpuscles—the appearances, indeed, which are described as characteristic of acute red softening of nervous matter.

Mr Stirling, Assistant Conservator of the Anatomical Museum of the University of Edinburgh, kindly undertook to harden some of the diseased portions of the cord, and to prepare some tinted sections. Mr Stirling hardened the cord by means of a mixture of alcohol and acetic acid, and attempted to colour the sections of it by means of an ammoniacal solution of carmine.

When a healthy spinal cord is submitted to this process, it is found that the pigment only partially adheres to it. The nerve cells, especially their nuclei, and the central portion of the nerve fibres, alone becoming coloured, the peripheral portion of the fibres (white substance of Schwann) remaining of the purest white. These differences in the absorption of carmine point to a chemical as well as a structural difference in these portions of the healthy cord.

In the diseased cord Mr Stirling observed, and by means of many preparations we have had the opportunity of verifying, that the carmine tinted the intercellular substance, the cells, and the different portions of the nerve fibres indiscriminately. The chemical differences, which in the healthy state exist in the different parts of the cord, appearing, as a result of the diseased action, to have disappeared. It would be interesting to know whether, in other cases of acute red softening, this peculiarity in the carmine reaction has been noticed.

The case narrated above appears to us to be one purely of acute red softening of the substance of the cord, *i.e.,* a softening the result of an acute congested and inflamed condition of the cord—a disease which has been called by some authors, Myelitis. The case differs from the majority of analogous ones in several particulars, especially in the sudden occurrence of paralysis. It is rare to witness this symptom at the outset of the affection. Usually the animal begins to walk stiffly, and to experience difficulty in bending at the loins. There is often evidence of pain in this region, especially when direct pressure is made on the

spinous processes of the vertebræ immediately above the
seat of disease.   After a period varying from two to three
days to some weeks, the animal becomes paraplegic, *i.e.*, the
posterior extremities become completely paralysed.   In the
worst cases, the animal is unable to void or retain urine, and
fæces voluntarily, and the posterior extremities often mortify.

It is not to be understood that such cases always follow
the course indicated.   Doubtless in many the stage of congestion is scarcely passed, and the structure of the cord is
not involved to any serious extent.   Were time allowed, in
many of these cases the animal would gradually improve,
and might, in the course of time, regain to a considerable
extent the lost power.   When, however, there is loss of
control over the bladder and rectum, the case must be
looked upon as so serious as to warrant, in almost every
case, the destruction of the animal.

Treatment should, in these cases, be directed particularly
to diminishing the congested condition of the cord.   In the
smaller animals cupping may be had recourse to with advantage.   Great attention must be paid to the condition
of the bowels, which have a great tendency to become constipated.   There are certain remedies which exert besides a
useful influence, probably by inducing contraction of the
minute vessels of the cord.   Of these belladonna and its
alkaloid, atropia, and the ergot of rye, are the chief.

The following prescriptions would answer very well for a
large dog :—

> ℞ Atropiæ grj.
> Acidi sulphurici dil. mins. x .
> Aquæ font. ℥viij.
> > Solve.
> *Liq.* 40 drops to be given twice daily.
> The dose to be increased after some days.

Or,

R Ext. ergotæ liq. (Ph. Britt.) 3ii.

Liq. 5 to 16 drops to be given thrice daily,

Nux vomica and strychnine are unfortunately often used indiscriminately in the treatment of all forms of paralysis. They must on no account be given in the affection now under consideration.

LOUPING ILL, OR HYDRO-RACHITIS IN SHEEP.

This disease is met with in many of the grazing districts in Scotland, prevailing in certain localities, and separated only by some arbitrary line from others immediately adjacent, where the malady is frequently almost entirely unknown. A remarkable instance of this is met with in the counties of Selkirk and Peebles, where, in the farms of the north side of the Tweed, the disease is very destructive in the spring and summer months, while on the south side of the river it is of extremely rare occurrence.

Professor Murray, of Cirencester, who investigated this disease in the summer of 1862, mentions that on the onset of the disease, " the animal falls down and struggles convulsively, paralysis has not yet set in, but the functions of the nervous system are disordered, the limbs are no longer subject to the control of the will, but plunge about convulsively." This is followed by a rapidly advancing paralysis of the limbs affected, which are most commonly the hind. The animal staggers with the fore or hind limbs, or towards the affected side, but soon loses all control over the diseased extremities, and is compelled to maintain a recumbent position. Being unable to move about in search of its food, the animal becomes weak and emaciated, and dies in a period varying from a few days, in the acute cases, to several months in the chronic. Coincidently with the onset of the paralysis, the appetite may be depraved, although it is usually rather voracious, and after death the stomach is often found to contain earthy matters and hair balls. In the advanced stages the urine and fæces are often discharged involuntarily, and in similar circumstances amaurosis and unconsciousness are not uncommon.

In *post-mortem* examinations, Professor Murray invariably met with increase of the cerebro-spinal fluid, while Mr Mathewson never met with this condition, but noticed in some that the cord was paler and softer than natural. The cases seen by the former were, however, more acute, extending only over a few days, while those seen by the latter seem to have extended over several weeks. The discrepancy may thus be partly accounted for—a part of the liquid having become absorbed in the interval. In some lambs killed shortly after the onset of this malady, we have found no great increased vascularity of the membranes covering the cord, and a thicker or more gelatinous condition of the cerebo-spinal fluid.

The causes of this malady are not sufficiently understood. They are doubtless intimately connected with the geological formation of the soil and the consequent modifications of the grasses 'and water on which the animals subsist. These causes are aggravated by exposure, sudden vicissitudes of heat and cold, and all such influences as reduce the general stamina of the animals. On this subject Mr Mathewson remarks—" It is curious also to notice the influence that cultivation sometimes has on this disease. On those southern exposed farms"—referring to those on the north bank of the Tweed—" I have already instanced that before they were cultivated to the extent they now are, and while mostly lying in rough pasture, the disease, I have been told, was not more common than on the opposite side of the river.   It is only since all the low-lying and rich land has been torn up by the plough and improved that it occurs to such a fatal extent.   Here I must explain, that since these improvements have been made on the land, and sufficient turnips and artificial grasses raised, the sheep are brought from the hills or unimproved portion of the farm, and folded on turnips during the winter, and are subsequently kept in the young grass fields till after the lambing season.   True, a good many lambs may die while in the parks, but the old sheep are seldom affected till they are driven to the hills or to the rough natural posture."   On some farms cattle and pigs are affected as well as the sheep, and the pastures can only be safely grazed by horses.   Mr Mathewson looks upon it as entirely due to coarse grasses grown on certain soils, adding, that " on some farms that have admitted of being entirely cultivated the disease has almost totally disappeared, lambs only being liable, and that to an inconsiderable extent."   On many farms it will attack animals in good condition, but in all cases thin and ill-conditioned animals form the majority of its victims.   Want of proper food, cold, and wet, are the chief exciting causes.

Prevention.—Keep the sheep constantly in good condition ; allow rock salt in covered troughs as a stomachic ; ameliorate the affected land by cultivation and by growing finer qualities of grass, and avoid putting sheep on such pastures as do not admit of improvement.

Treatment.—This is of course very difficult when a large number are affected.   Mr Mathewson succeeded in curing three out of five, by supporting the strength with hay, cut grass, and gruel, and by exhibiting for a week diuretic doses of nitre, digitalis, and oil of turpentine, followed by a course of nux vomica.

TREMBLING

Is a somewhat badly defined disease. The term is some-times applied by shepherds to almost any internal inflam-mation, the onset of which is characterised by a severe shivering fit. It is sometimes applied to a form of louping ill, when that is attended by marked muscular tremblings

THORTER ILL.

This is a parasitic disease, in which the hydatid is situated in the cervical portion of the spinal cord, and is attended by more or less paralysis of one or both sides of the body.

HEMIPLEGIA

Is a rare affection in the lower animals, and is commonly dependent on effusion of blood on one side of the brain or some lesion of one-half of the spinal cord. According to the cause, it may affect one entire half of the body, in-cluding the head and even the intercostal muscles, or i may be confined to particular portions of one side of the body, and especially such as derive their nerves from tha part of the spinal cord situated behind where the morbic lesion exists.

CANCER OF THE SPINE.

A remarkable case of this description was published in the Veterinarian, 1856, by Mr Hunting, with note of my own as to the cadaveric lesions. Mr Hunting says :—

" On the 10th of September 1855, I was requested to see a chestnu mare, the property of Mr George Reed, of Seaham Harbour. She ha been unwell for ten or twelve days, with cough and sore throat ; he neck was likewise very stiff, but her appetite had remained good up t yesterday, when she became tympanitic, and suffered intense pain.

" When I saw my patient, I found that the pulse numbered 68, and was weak. The submaxillary artery appeared full and soft, and the action of the heart feeble. The breathing 59 in the minute, and rather laboured. The conjunctival membranes were very much injected and of a yellowish colour. The mouth was hot and dry. The nostrils were greatly dilated. The ears were cold, legs warm. The fæces were of a healthy character. The respiratory murmur throughout the whole length of the trachea was much louder than in health, and the left lung gave evidence of partial congestion. The cough was thick and heavy, and of a peculiar sound, but not frequent. The larynx and trachea were very painful on pressure. A watery fluid flowed from the eyes, but there was no discharge from the nostrils. The neck was very stiff, so much so that the animal could neither eat nor drink from off the ground, nor move the head in a lateral direction. The parotid glands were much larger and harder than usual. The appetite, however, was but little impaired.

" From these symptoms, I considered it to be a case of ordinary influenza, or distemper, which disease was exceedingly prevalent in the neighbourhood at the time. Acting upon this impression, I treated it as such until the 24th, when my patient was so much better that professional attendance was no longer necessary. The head could now be moved with greater freedom in a lateral direction, and the mare was enabled to eat and drink from the ground. The breathing had become natural both in character and frequency. The pulse 38, and healthy in tone ; the appetite good ; the animal lively, and capable of taking half an hour's exercise daily.

" On the 6th of October I received a message to say that the mare was not so well. On arriving at the place late in the day, I found her apparently suffering but little pain, the intense agony which the owner had observed in the morning having passed off. The other symptoms they described as also existing were an enormously distended abdomen, frequent groaning, rigid limbs, an occasional lying down but quickly rising again, very heavy breathing, and an anxious expression of the eyes.

" The pulse was 44 in number, and rather weak at the jaw ; the sounds of the heart were so feeble that they were scarcely audible on the left side ; the breathing was 66 in the minute, but not laboured ; the motion of the abdominal muscles was indeed so slight that I was unable to take the number of respirations at the flank ; the neck was still a little stiff ; the nostrils dilated, and much anxiety of the countenance present. The mucous membranes were healthy in colour; the mouth cool and moist; the surface of the body of a natural temperature; the bowels regular, and the urine of a light colour.

" On the application of pressure to any part of the spinal region rom about the tenth dorsal vertebra to the sacrum, the whole of the voluntary muscles behind became as rigid and hard as in the worst cases of tetanus during the periods of excitement. Very little difference could be detected in the violence of the muscular contractions, whether the pressure was employed directly over the spine, or within twelve inches on either side of it. This tetanic rigidity sometimes occurred when pressure was not applied, and also when the animal was made to back, but then in a much less degree, and lasting for a few minutes only.

" On an examination per rectum, I detected a large tumour on the left side of the spine, having a density, as imparted to the feel, equal to the structure of the liver. It appeared to be about eight or nine inches in diameter, three inches thick in its centre portion, and an inch at its circumference. It was closely connected to the posterior part of the kidney, overlapping the posterior aorta, and extending to the right side of the spine. On pressing the enlargement, evidence of severe suffering was obtained, but the tetanic spasm did not follow, nor was there the slightest indication of pain when the pressure was applied to the inferior portion of the lumbar vertebræ. The pulsation of the posterior aorta between the tumour and the bifurcation of the vessel into the iliac. arteries, was scarcely to be felt, and which I attributed to the pressure of the enlarged mass upon the aorta.

" I looked upon the case as one very doubtful of recovery, thinking that I had most likely to deal with the formation of an internal abscess, as a sequela of influenza, and which is not a very uncommon occurrence; but as my patient's appetite was good, the heart's action not much disturbed, and the fæces and urine healthy, I considered treatment justifiable. Counter irritants were therefore applied to the loins, and Pot. Iodidum given internally, with vegetable tonics. Under this treatment the tumour gradually became less in size and much softer in consistence. The rigidity of the muscles was likewise less violent when pressure was applied to the spine.

" On the 24th, a ' charge' was applied to the whole of the lumbar region. Mineral and vegetable tonics were daily given with the iodide of potassium, and exercise was enjoined.

" From this date up to the 18th of November, there was a gradual wasting away of the muscles; the appetite was generally good, the animal eating an average quantity of the most nutritious provender that could be procured.

" During this period the urine became highly impregnated with albumen, but which gradually diminished in quantity, until on the 18th of November, after which time I entirely failed to detect its presence.

" On several occasions during the above period, the peculiar spasmodic contraction of the muscles, the tympanitic state of the abdomen, the anxious and protruding eyes, and the intense suffering, would come on, but only lasted for a short time. These symptoms generally occurred two or three days in succession, and were not again seen for six or eight days.

" On the 18th of November, I found that the tumour was very much smaller, and also soft and flaccid. The pulse and breathing were perfectly natural, both in character and number. The spasmodic contraction of the muscles did not occur, even when pressure was applied to the spine, and the animal looked more cheerful and lively.

" From this date up to the beginning of December the mare slightly improved in condition, and I had more hopes of her ultimate recovery; but on the 6th of December the symptoms again returned, and in as bad a form as before. They continued for several days, and then disappeared.

" On the 23d of December, at 8 o'clock in the evening, she was left apparently no worse. She drank an unusual quantity of water, and ate her food with avidity; but early on the following morning she was found dead and cold.

" *Post-mortem appearances.*—The thoracic viscera, the spleen, and the liver, were all very pale in colour, but otherwise healthy. The left kidney was very much enlarged. The stomach, intestines, bladder, uterus, and right kidney, were likewise healthy; but the pancreas was filled with small tumours, varying in size from a pea to a walnut, which contained a yellowish jelly-like substance. The spinous processes of the lumbar vertebræ were extensively diseased.

" The second, third, and fourth cervical vertebræ were extensively diseased. In the broad, flat, spinous process of the dentata, there existed a circular aperture, extending from side to side, and from the arch to the top of the process at the point of its bifurcation. On the lateral and anterior part of the body on the left side the disease had made nearly equal ravages, extending into the foramen at the base of the odontoid process. Immediately above the transverse process, on the same side, the destruction of bone extended from an inch and a half upwards and backwards, invading a nearly circular portion, and removing a large part of the articulation formed by the left half of the spinous process of the dentata and the anterior articular process of the third vertebra. At the antero-inferior part of the body of the bone the disease had established a complete communication—nearly an inch in diameter—with the spinal canal, but the dura mater was not destroyed.

" In all the affected portions there was a considerable quantity of bony material removed, forming large cavities, which were filled with a reddish-looking mass, and presenting precisely the same appearances as those of the lumbar spine. In the third and fourth vertebræ the lesions were less extensive, but showed the same characteristic appearances."

In a letter addressed by me to Mr Hunting in the month of March 1863, I gave the following description of the morbid appearances :—

" The kidney is greatly enlarged, weighing thirty-six ounces, flabby, of normal colour throughout, and its pelvis contains a large quantity of epithelium in a scanty fluid.

" Connected with the posterior part of the kidney is a portion of aorta, around which the cellular tissue forms circumscribed cysts by its condensation, and is infiltrated with pus. The pus is homogeneous, creamy, strongly charged with corpuscles, whether nucleated or simply granular, and not mixed with heterologous productions. I have failed to determine any relation between the abscesses in question and the lumbar glands ; though, on cutting into the mass at first, a gelatinous infiltration caused me to suspect I had to deal with suppurating lymphatic glands.

" With reference to the portion of spine consisting of five lumbar vertebræ, the bodies of two contiguous ones are broken down and destroyed : they are perforated from side to side, and the cavities formed contain bony spicula, in great part held together by the inferior vertebral ligament, and imbedded in a violet-red pultaceous mass of the consistence of brain substance. The spinous processes of the first two appear healthy ; the second slightly hypertrophied at its base. On the third, which is the one with extreme disease of the body, the bone structure at the base of the spinous process bulges considerably on either side. On cutting into this, on the left, is a clear, semi-transparent, bluish-gray deposit, of gelatinous consistence, with no evidence whatever of inflammatory action around, but clearly malignant in its nature, and surrounded by expanded bones, which has made way for the morbid product. The spine of the fourth vertebra is similarly affected, but to a greater extent. The transverse processes have been equally invaded. The second one presents, on its superior surface, towards its attached end, a circular aperture, with tolerably defined margin, communicating with a cavity filled with malignant deposit, and circumscribed by the expanded layers of the bony process. The last two of the spinous processes are much more diseased, the rarefaction of bone is greater, the destruction more advanced, and a jagged erosion, about an inch long, and more than half an inch in extreme breadth, indicates where the cerebriform deposit has spread its ravages rapidly and with effect. On the right side it is only the transverse process of the last lumbar spine, with the corresponding portions of the sacrum sawn off with it, that give evidence of diseased action. The bone, greatly rarefied and swollen, though looking solid throughout, when cut across is found to contain an irregular deposit of yellowish-grey colour and gelatinous firmness, which has led to a considerable excavation of the bone.

" On cutting through the arches of the spine on the left side, then perpendicularly through the bodies from below, the spinal canal is exposed without being injured, and the progress of the disease can be readily studied. In the centre of the cancellous tissue of the bodies of the last two vertebræ is a cancerous deposit, most extensive in the very last bone. It is seen in the form of perfectly circumscribed masses, which we must attribute to the colloid variety, from its transparency, uniform gelatinous consistency, and scantiness of cells. Not so, however, with the large pulpy masses, infiltrating and destroying almost entirely the second and third lumbar spines ; this is distinctly encephalòid, of apparently rapid formation, deeply tinged, of a brownish red in parts, or else of a yellowish or greyish red in others. The bone around, the solidity of which has given way to malignant disease, gives no evidence of inflammatory action, or any reparative process ; it has yielded to the encroaching deposit, which has forced two open passages into the spinal canal. Here blood has been extravasated outside the dura mater, pressing, as well as the carcinomatous mass, on several of the organs of

the lumbar nerves ; otherwise the cord itself does not seem to have bee
much compressed, and is perfectly healthy in structure.

" Histologically, a fact of considerable interest is the almost com
plete absence of cancer-cells in all the deposit.   Some nuclei, granula
matter, and detritus, is all I can discover.   Whether the cells have no
had time to form in the translucid gelatinous deposits, or have bee:
broken up and disintegrated in the encephaloid masses, it is difficult t
assert.   That, however, the morbid appearances above described ca:
alone be attributed to carcinoma may be inferred from the multiplicit
of the deposits.   Hearing that there was a rigidity observable at th
upper part of the neck, I was glad to see my expectations come true ii
your note of the 5th inst. ; after having taken the trouble to disinter th
skeleton of the mare, you discovered disease in every way similar t
that of the lumbar spines affecting the second, third, and fourth cervi
cal vertebræ.

" Cadaveric inspection, moreover, revealed that form of deposit whicl
generally occurs in the shape of circumscribed masses, or of infiltratior
within bone, viz., colloid disease and medullary cancer ; the latter witl
effusion of blood, hence its soft consistence and blood-stained, or hæma
toid, appearance.   No inflammatory deposit, no suppuration, existec
in the osseous texture ; and we must notice the deadly tendency of the
growth, there being no attempts to limit it, but infiltrating, invading
destroying, and transforming to its own nature all in its vicinity, perforating the spinal canal, unremittingly progressing to the annihilation oi
life.   These characters alone are sufficient to prove the malignancy of the
deposit, however scanty the evidence of the existence of cancer cells."

NEURITIS, or inflammation of a nerve or its sheath, may
be said to be unknown in the lower animals, unless as
occurring after neurotomy, when the divided end of the
nerve has been subjected to some source of irritation.

NEUROMA, or tumour on a nerve, is equally rare, though,
after the above named operation, a fusiform enlargement is
sometimes met with on the divided end of the metacarpal
nerve.   It has a fibrous texture.   If subject to continued
inflammation and pain, the tumour may be removed by
excision.

CHAPTER XIX.

ON THE FOOT AND THE ART OF SHOEING.*

Horn.—Secreting structures. — Papillæ. — Laminæ.—Horn cells. — Horn fibres.—Growth of human nail and horse's hoof.—On shoeing.—History of the art.—The wall, sole, and frog of a horse's foot.—Bones of the foot.—Peculiarities of the coffin-bone.—Prevailing errors on the subject. —Preparation of the horse's foot for shoeing.—The "rogne pied" or toeing knife.—French system of forging shoes.—Art of shoeing in England.—How to make shoes.—Application of machinery.—Form of nails. —The French method of forging shoes.—English shoeing.—Stamping and fullering shoes.—Comparison between English and French nails.— Oriental method of shoeing.—Comparison between English and French shoeing.—Relative labour in making fullered and stamped shoes.—Fullered shoe sometimes specially advantageous.—On the weight of shoes.— Number of nails. — Position of nail holes.—Toe-pieces. — On fitting the shoe.

HORN.

THE horns, claws, nails, and hoofs of animals are all composed of material similar to hair, and are often spoken of as built up of hairs firmly matted together. The same cell which forms the scaly epithelium, epidermis, and hair, is utilized in building up the horny structures. The special history of the horny appendages of animals consists, therefore, in the description of the form and disposition of the surfaces from which they spring.

* For assistance in the preparation of Chapters XIX. and XX., I am specially indebted to my father, who has prosecuted many inquiries into the physiology and pathology of horses' limbs and feet, and has thereby mastered the difficulties of the art of shoeing.

Whereas hairs have a root imbedded in a follicle, horn springs from papillæ, which stud a surface extended over a bony or fibro-elastic prominence. Thus the papillated tissue, whence spring the horns of cattle, sheep, &c., forms a covering to the processes of the frontal bones, which are pierced by large foramina for the transit of blood-vessels.

In the foot of the horse we observe the skin, at the part where hair and horn meet, thickened and altered in character, constituting the structure called the coronary band, and towards the posterior part of the foot the subcutaneous tissues consist of that vast fibro-cartilaginous cushion, constituting the elastic basis over which the resilient horn of the frog is formed. From the coronary band downwards, permanent folds, laminæ, or podophylla (Clark), are arranged in parallel lines. They are about 600 in number, and like the papillæ, destined for secretion of the agglutinating cells, which form the matrix of horn. As Virchow mentions with regard to the human nail, each lamina corresponds to the

Fig. 216.—(CHAUVEAU).—Different forms of horny scales. The cell to the left is one from the softer and deeper layers, and is charged with two pigment masses, × 300.

single papilla seen on the surface of the skin. The cutaneous surface beneath the os pedis, over the fibro-elastic frog and the lateral cartilages, forms, with the coronary band, an extensive bed of papillæ.

The surface of the horn has a fibrous appearance, and the

fibres run in a parallel direction, and in a straight line from the papillæ which form them. Thus the papillæ on the coronary band, frog, and vascular sole, are all directed downwards and forwards.

If a hoof is cut through perpendicularly, it is found that the deeper layers are soft, and the tissue becomes progressively harder from within outwards, so that the surface is firm

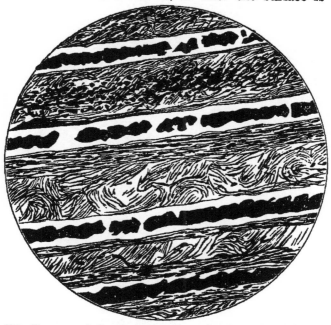

Fig. 217.—(CHAUVEAU.)—Longitudinal section of four horn fibres of the wall, taken from the point of union between the white and dark horn. The dark material in the centre of the fibres is composed of opaque spherical cells, which appear dark when seen by transmitted light.

and of a character suited to its functions, viz., for sustaining weight, maintaining a given form, and defending from injury; at the same time, it is endowed with elasticity, one of the essential properties of hoofs.

The so-called horny fibres are funnel-shaped at the point of connection with their respective papillæ. Each fibre has

a medulla or pith, composed of soft spherical or polyhedra
cells, and of a cortical portion, which has a fibrous appearance, due to elongated fusiform cells, like the cortical portion of the hair.   The thickness of the fibres varies; those at
the surface of the wall of a horse's hoof are the smallest and

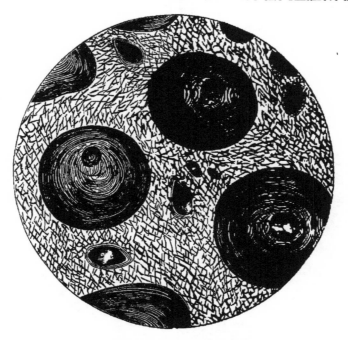

Fig. 218.—(CHAUVEAU.)—Transverse section of several horn fibres of the horse's sole,
showing the concentric arrangement of scales around a central medulla; also the super-
imposed cells constituting the agglutinating material, × 100.

most compact.   Between these fibres there is an intermediate
substance, best seen on examining a transverse section of
horn.   Such a section (fig. 218), when examined by transmitted light, affords us a view of the concentric layers of
horn-cells surrounding the soft opaque medulla, besides the
connecting substance formed by a dense mass of cells similar
to those which constitute the fibres, but disposed in an oppo-

site sense, lying flat on each other at right angles with the fibres. This dense bed of cells forms the bond of union between the fibres, which become loose and detached from each other when the horn by accumulation is removed from its proximity to the source of production, and its integral strength is lost.

The fibres, we have said, are formed from the papillæ, but the cells of the material which joins them are produced from the surface between the papillæ and the deeper layers from the surface of the laminæ or podophylla. Thus the horn fibres descend over the laminæ, and are attached to these by cells constantly developed, and which act as the agglutinating material.

The pigment which colours all dark horn is disposed irregularly between the cells of the fibres, as in the case of hair, and has no very regular disposition.

The difference between the horn of different parts of the horse's hoof depends not a little on the relative amount of fibres and inter-fibrous substance. The horn of the wall is tough, and breaks up into fibres as it grows beyond a natural length. This is due to the toughness acquired by the fibres in their lengthy course, and to the crumbling of the cells between them as mentioned above. The horn of the sole is detached in flakes, and this depends on the fact that the fibres are short, do not pass over or into a bed of agglutinating material, and when at a certain distance from the papillæ whence they originate, they break off. Flakes are thus detached from the frog. The growth of horn is unlimited, and in the case of the horse's hoof, we find that if, from the feet being protected by an iron shoe, when injudiciously managed, whether through want of exercise and proper ground to move on, certain parts are allowed to grow beyond the natural depth which would not occur if the animal was free in a state of nature, they imprison other parts between them, and lead to the accumulation of masses of firm unyielding horn, which inflict injury. The unlimited growth of the fibres is beautifully illustrated by cases in which the hoof is not worn down owing to an animal's limb being distorted, and the hoof brought to the ground on its side instead of its normal surface of apposition.

The growth of the wall of a horse's hoof is in every respect similar to the growth of a human nail. We shall quote from Virchow:—" If we consider the nail with respect to its proper firm substance, its compact *body* (Nagelblatt), this only grows from behind, and is pushed forward over the surface of the so-called *bed of the nail* (Nagelbett), but this in its turn also produces a definite quantity of cellular elements, which are to be regarded as the equivalents of an epidermic layer. On making a section through the middle of a nail, we come, most externally, to the layer of nail which has grown from behind, next to the substance which has been secreted by the bed of the nail, then to the rete malpighii, and lastly to the ridges upon which the nail rests.

"Thus the nail lies in a certain measure loose, and can easily move forwards, pushing itself over a moveable substratum, whilst it is kept in place by the ridges with which its bed is beset. When a section is made transversely through a nail, we see, as already mentioned, essentially the same appearance presented as that offered by the skin, only that a long ridge corresponds to every single papilla seen in ordinary sections of the skin; the undermost part of the nail has slight indentations corresponding to these ridges, so that, while gliding along over them, it can execute lateral movements only within certain limits. In this manner, the body of the nail which grows from behind moves forward over a *cushion* of loose epidermic substance in grooves which are provided by the ridges and furrows of the nail. The upper-most part of the nail, if examined when fresh, is composed of so dense a substance that it is scarcely possible to distinguish individual cells in it without applying reagents, and at many points an appearance is presented like that which we see in cartilage. But by treating it with potash, we can convince ourselves that this substance is composed of nothing but epidermis-cells."

As Virchow justly observes:—" From this mode of development, an easily intelligible distinction may be drawn between the different diseases of the nails," and we might add, of the horse's hoof. The causes of false quarter and seedy-toe, the appearances of canker, are most satisfactorily explained. There is only one of these, affording a useful illustration of the manner in which horn is formed, that we shall rest upon. It is the deformity resulting from the diseased condition hitherto called laminitis,—Knollhuf of the Germans,—in which a great enlargement occurs at the toe, believed generally to be due to a descent of the os pedis. The change is gradual. From inflammatory action, a separation occurs between the podophylla (vascular laminæ), and keraphylla (horny laminæ.) As the inflammation subsides, the space, however small, becomes filled up by the cells which usually agglutinate the horny fibres; but as these are detached, they only become more and more elevated, and the space between the horny wall and the os pedis increases. The toe of the latter becomes atrophied, and it is impossible to obtain a restoration of the wall, because the fibres are pushed outwards by the enormous mass of cells found beneath them.

There is a disease of the human nail (Onychogryphosis), in every way similar to the above-mentioned deformity of the horse's hoof. Virchow refers to it as follows:—" When there is a very abundant development of cells in the bed of the nail, the body may be pushed upwards, nay, it sometimes happens that the nail, instead of growing horizontally, shoots perpendicularly upwards, the space underneath being filled with a thick accumulation of the loose cushiony substance."

On Shoeing.

On the high claims to the veterinarian's attention of this department of his calling, little need be urged in this work, since most writers, and almost all able men who have in any way advanced the art of managing horses, have amply set forth the requirement of a rational system of shoeing to be generally carried out in practice.

It is the inconsistency between that which is very generally acknowledged to be an essential requirement, and the indefinite diversities of prevailing customs, which prompts us to take up the matter with becoming earnestness, as the one of all others which interests a large proportion of readers.

History of the Art of Shoeing.

It would be going beyond our limited bounds in this place, to attempt any lengthened historical account of the subject. Still it is believed that nothing tends more to establish knowledge of a science, or an art, than the tracing its past history, each earnest worker thereby fortifying his understanding on the means by which advances have been achieved, the influences which have tended to hinder, and the causes of failure, where such has happened.

Man, unlike every living creature besides, works to-day by the light of past ages, and owes to his fellows, and to future generations, the obligation to use his talents to the utmost in forwarding the cause of truth in every thing he sion, act by instinct; the bee of to-day following the same laws as those of his species in all preceding generations.

The history of any subject which extends back for ages, is always difficult to unravel, and when sought to be gone into, is usually lost in remote obscurity; such is pre-eminently the case with the history of horse-shoeing.

The late Mr Bracy Clark applied his classical learning, and great love for the subject, with much earnestness, and after all his researches, believed that horse-shoeing had been in vogue for twelve or thirteen hundred years, and on the credit of some traditional accounts, speaks of its having been brought into Britain with William the Conqueror.

It appears to us that we are totally unable to fix on any date, even approximately, on the origin of shoeing. We have no account whatever of the beginning; unlike the case of the art of printing, we find no name attached as that of the original inventor; and though we are instituting inquiries into an appliance which has enabled man to avail himself of the horse, as a means of advancing civilization, beyond any other power which he could control; and when, in recent times, other powers are made to supply those of the horse to a great extent, that animal is brought into even increased requirement, and, like man himself, does much labour which only living locomotive powers can effectually perform. Notwithstanding all these reasons, there is no epoch to which any two authorities have agreed to assign as that to which the art in question begins to take its date.

We say, then, with Berenger, in his investigations into the history of the horse and horsemanship, that the very absence of any recorded incident whereby a date can be fixed on, is proof of a high antiquity; and, in truth, it must be confessed, that we have no account whatever of the origin of horse-shoeing, but are in total ignorance even of the nation in which the art was first applied, and equally so as to the quarter of the globe in which the ingenuity of man was brought to bear on the subject.

Nor is the negative side of the question more instructive; local accounts must not be taken as full evidence, such as the absence of any account of shoeing horses in the army which Xenophon commanded, or the non-appearance of shoes on the equestrian works of the early sculptors, because the same countries whence these accounts come, admit of horses being used to a great extent at the present day without shoes; therefore, the history of horse-shoeing, as far as it can be made instructive to the many, lies near to our own time; from the latter end of the sixteenth century to the present period may be found all that is necessary to show the state of the art, at the epoch referred to, when Solleysell wrote, proving himself by his work to have been the first and most able of all modern authorities on the subject. The art was, no doubt, well advanced at that time in Spain and Italy, and we believe is referable to a much higher antiquity in the old world, about which we have no early accounts, and yet, looking at the methods of horse-shoeing adopted now amongst eastern nations where no European has had any part in effecting change, evidence is afforded of an innate intelligence.

In all the specimens of shoeing that we have seen, from the remotest countries from which travellers have brought them, however rude the workmanship, a clear intelligence is evinced in adapting means to the requirement.

This subject embraces a wider field for combined labours than may be apparent under its title; to approach proficiency, a clear understanding of the locomotive functions of the horse is required, and if that be acknowledged, the necessary steps must be taken, by going back to the ole

ments of anatomical science. The practice of farriery, viz., that branch of the veterinary art which takes under its charge the art of shoeing and the treatment of lameness —subjects intimately allied and inseparably connected—in which, as a whole, science, artistic skill, and physical activity, are called into requisition, demands the resources of self-denying, able men.

Whilst showing that it is no common smith's work that we are engaged to give an exposition of, it may be readily seen that it is a work for many hands, and does not admit of being equally cultivated to perfection as a whole by all who take it up. Let the same division of labour be encouraged in the art of horse-shoeing as prevails in every other, and something will be done towards arriving at a better general knowledge, and instead of the matter being looked on as everybody's business, whilst no one thinks it incumbent on himself to take the necessary pains to master its details, each man will be induced, it is hoped, to do something in its furtherance, and share in the honours and responsibilities accruing.

The relative share that different workers may take in the cultivation and practice of shoeing is not readily distinguishable in any marked degree, if the hand has need of the head, so does the latter depend on the former.

The physiologist may be supposed the most likely to open up new grounds, whilst the ingenious worker, applying his resources of art, with the help of some correct knowledge on the movements of the foot, will so accommodate means to requirements, that the art will contribute in turn its share to the science.

In the range of knowledge necessary for the prosecution of the art of farriery, a body of disciplined men are required, possessing abilities varied in kind, which can only be found in the many, amongst whom every phase of the subject might be expected to have its adherent.

The whole foot of the horse should be viewed in its proper aspect, embracing structures above the hoof as well as those enveloped by it ; details may advantageously precede general arrangements, in which way the separate parts of the foot should be investigated, they being to the physiologist what the letters of the alphabet are to the scholar—the first steps of the ladder, the parts are learnt separately, and then blended as the understanding puts the knowledge into useful form.

Nature has furnished the horse with hoofs, which are endowed with given degrees of substance, density, toughness, and elastic properties, differing in their different parts, so as to assimilate the functions of the outer with the inner structures.   On that animal, more than others, is conferred a double framework to the foot, an inner bony skeleton and an outer encasement of horn, and these mutually act in sustaining and supporting the great exertions which the foot, as the support and lever, undergoes.

Subordinate to the sustaining structures just referred to, those more pliable are brought into their assigned action, viz., the ligaments and cartilages, amongst which may be comprised that immensely powerful structure, the frog, with the intercurrent ligamentous bands which take their attachments from the lateral cartilages, and converge to the centre; this anatomical formation is shown, and the functions described in the *Edinburgh Veterinary Review* for 1861, at page 511, though a single illustration, however well executed from nature, is totally inadequate to do more than aid the text, to indicate how each investigator may succeed in finding out the parts by his own subsequent dissections.

The hoof of the horse may be described according to its

formation, in three distinct parts, called respectively the *wall*, *sole*, and *frog*.

Each division of the hoof is composed of horn, differing in texture at different parts, that which enters into the composition of the wall is denser than that of the other divisions, it is of a fibrous nature—the fibres taking a longi-

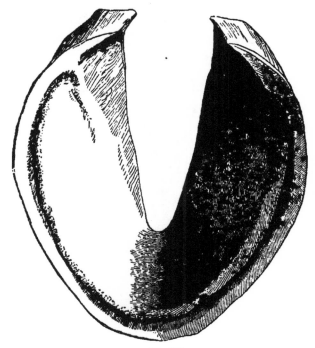

Fig. 219.

tudinal direction from the upper part or coronet downwards, and are endowed with an elasticity and density which admirably fits them to embrace the complex structure, support weight, and resist wear. The density of the horny fibres is greatest, as we have said, as they approach the surface of the wall, the outer layers acting as a cuticular covering to defend the inner from external agencies, snch as a dry atmosphere, excessive moisture, &c., performing, in fact, the same functions as the cuticular covering does to the true skin of all animals over the whole body.

The depth and strength of the wall are greatest at the point or toe. Posteriorly, on each side, it is inflected inwards, so as to form an internal wall. These inflections have been described as separate organs, and named the *bars*, and accordingly, a function has been erroneously ascribed to them, viz., that of propping open the heels.

Fig. 220.—Frog of the horse, showing the separation of flakes.

The sole is the next division of the hoof to be noticed, and viewed in connection with the wall, it may be regarded as the arched support of the foot; it is an irregular thick plate of horn, presenting to the ground a more or less concave surface; its outer margin furnishes a broad surface for attachment to the inside of this part of the wall, where it blends with the horn plates.

The frog is a triangle of very elastic horn, filling up a space of corresponding figure in the sole; it extends nearly two-thirds through the centre and bottom surface of the foot. At its anterior part or apex, it consists of one ridge; about one-third of its length from the point, the organ divides into two equal parts, leaving an interspace termed the *cleft*. By this arrangement, provision is made for motion in the posterior parts of the hoof.

The frog is composed of horn of a fine tough texture, more elastic and pliable than the sole; besides, it covers an exquisitely elastic organized structure of its own shape.

Some notice of the bones of the foot may now be taken; on their form, relative position, and connection depends its motion, which in the horse, physiologically considered, begins where the radius terminates in the fore, and the tibia in the hind limbs, that is, the knee and hock and all below enter into the pedal function.

The bones which enter into the construction of the foot should be understood with reference to their particular functions; to the required action of these solid parts, all other structures concerned in the locomotive functions are made subservient.

The bones of the limbs belong to two distinct orders as regards function, one of which, sustaining the weight, are acted on, and are the levers which give velocity, and all motion; the others are formed into projecting pulleys, buttresses, and also become levers in connection with the shafts alluded to in the first order, of which the cannon, pastern, coronary, and pedal bones, constitute those of the foot extremity in both fore and hind limb, and of the latter, there are two pairs, and two single bones in each extremity, viz., the navicular, the two sessamoids, and the two splint bones, with the pisiform in the fore, and the os calcis in the hind limbs; in this classification, the other bones special to the knee and hock are not taken account of

The function of the bones of the last order described is necessary to be understood, inasmuch as it differs from that of the sustainers, and it is only by understanding these in detail that anything can be fully estimated, either of normal action, or of disordered conditions.

The bones which are all placed behind the axis of bearing, constitute the medium by which muscular energy is made to act with great force, but these have little more to do in sustaining the superincumbent weight, than have the trochanters of the femur, the use of which is obvious to all anatomists, viz., that they form projections for uplifting power; taking another example, we may adduce the patella, the great and essential offices of which are most distinct from that of sustaining the superincumbent weight; and just in the same way may the navicular and sessamoid bones be viewed; whilst the splint bones, and those projecting posteriorly of the knee and hock, are so fixed and acted on directly by muscular power, that they constitute additional parts of the shafts, a main function of which is their uplifting action.

The os pedis, or coffin bone, is peculiar in the horse, both in its structure and economy; there is a close analogy between that bone in the horse, and the double formation of the same in the ox, but the resemblance is only partial, each animal being perfect for the uses and situations for which it was designed; the cloven-footed animal moves with astonishing security over granite rocks, where the horse is less adapted to go; this fact is illustrated in the different kinds of goat, deer, and in a lesser degree in the ox. All cloven-footed animals are endowed with remarkable security of foot-hold, but want the elasticity to carry weight and the graceful

dale, the situations suited to that animal's whole nature, where he finds sustenance for life, and where alone his powers, speed, and endurance are required.

The pedal bone has much of the form of the hoof in its

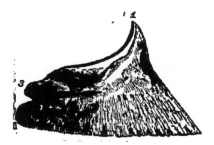

Fig. 221.

exterior aspect; and when the cartilages, with the other structures which attach to the bone, are seen in connection, with the coronary and navicular bones in position, the whole organized structure is similar in its outermost form to that of the hoof.

As has been said, we cannot, if we wish to understand the subject, confine our views to that which is generally treated as the horse's foot, viz., the hoof, and the organs it envelopes, but a larger understanding of the structure of the limbs and the locomotive functions generally is necessary.

The horse's hoof is not to be regarded simply as a covering for the protection from external injuries of other structures, namely, that of the sensitive parts. The hoof has its specially assigned sphere in the whole economy of the foot, and each separate component part must be looked on as an essential constituent of the whole organization. Nature does not make one part so imperfect as to require another to minister to its aid.

The hoof forms an integral part of the foot, and those

animals that lack it, though amply protected as they are, cannot sustain weight and undergo the same fatigue on the same
soil as those which possess it; and of the animals gifted with
hoofs, the horse is superior to them all.

The hoof must be studied in reference to its construction
and economy as the index to the development, condition, and
health of structures beneath it, standing in relation as it
does to these parts the same as the skin does with the structures it covers.

The horse's hoof is commonly regarded as a secondary
structure, a something that may be cut away to ascertain
the condition of parts under it, not understanding the fact,
that when such exploration has been perpetrated, the normal condition of the whole foot is interfered with for many
months afterwards. If a medical man examines the limbs of
a man, he is thoroughly cognizant of the health or otherwise
of the whole structure by the condition of the skin, and so
should the veterinary surgeon be with regard to the horse's
foot, by taking the hoof especially for his chief guidance;
the organs of vision and touch being fully adequate to furnish the mind with data, when to the rest the horse's action
is taken into account. When the questions bearing on the
above phenomena are satisfactorily solved, it will readily be
admitted that the hoof is a most important structure for its
own particular part in the whole economy of the foot, and
that it must be maintained in its duly proportionate form,
with its natural density, whence comes elasticity, and strength
to sustain weight, and maintain its required form.

In describing the bones of the horse's foot, whilst limiting our observations according to the scope of this essay,
there are only three which come under our notice. These
are, first, the coronet bone, the foot or coffin bone, and the

is formed—a joint exquisitely beautiful, and of the first importance in its exactly assigned functions.

The coronary bone forms the medium through which all the weight is exerted, which is conveyed to it by the pastern bone, and which is lastly diffused through the coffin bone.

The action of the coronary bone is most considerable at its lower extremity, where it has a large articulating surface in connection with the coffin and navicular bones; its action in progression is a revolving movement from the posterior on to the anterior surface of its lower condyles.

The coffin bone is the last in the limb, of those in the axis of bearing, through which all power is exerted, and by the mode of connection between it and the hoof, great weight is carried, with velocity unexampled in other animals, and is distributed on to the ground with the most perfect freedom from jar, whilst the inherent strength in the whole foot is prodigious.

On reference to the pedal bone, as shown in the foregoing illustration (Fig. 221), it is seen to resemble, in its leading features, the external form of the hoof; one chief difference being found to consist in the former being fully a fourth shorter than the inner cavity of the latter. When, however, this wonderfully constructed bone is furnished with the cartilages, ligaments, tendinous attachments, and all the influential structures of which it constitutes the centre, the entire concentration of powerfully organized parts, assume the form, and become the counterparts of the hoof.

Though it is not the aim on this occasion to supply a great present want, viz., that of a treatise on the economy of the horse's foot, with ample details on the art of shoeing, it is none the less hoped that, to the extent to which this article goes, little of what is given will be set aside hereafter, or require to be withdrawn.

It is believed that, much as principles are required to be laid down and spread on all that relates to horse-shoeing and the management of horses' feet, that in this place good will be derived by our trying to make the subject plain for popular reading, rather than elaborate. A first move in a right direction, constitutes the extent of our aim in this work.

With reference to the organization of the foot and its functions, so far as these essential details are discussed, it will be sought to make the inner movements apparent from an external view, as most of the phenomena which form links in the chain of facts, the understanding of which is necessary for the establishment of a successful practice of shoeing, may be seen in the way alluded to.

We are of opinion that the ancients drew their conclusions on the movements of the horse from observing his external appearances chiefly, and most of what is best understood on the horse's foot at the present time has been acquired by men who have observed carefully, and at the same time handled the parts contemplated most industriously; thus making the two gates of knowledge which constitute the most effectual channels, the rational means for the attainment of an end.

While contemplating the foot with reference to the adaptation of systematic rules on shoeing, we must take account of the innermost structures, find out the course of bearing as it is transmitted from the longer shafts to the coronary and pedal bones, and how this last disperses the weight over the hoof, and through it to the ground; we find, as is the case from the knee and hock downwards, that at the extreme point the greater solidity exists in front, whilst the structures are all yielding posteriorly; the cartilages on either side commence where the extensor tendon and the branches from the carpo-pedal ligament are inserted into the prominent pyramidal process of the coffin bone; these may then be traced backwards where they are strongly connected to the basilar process, and where much increase of substance is found with more regularity in form; the lateral cartilages have a depth, and from a clear space above the hoof reach to the bottom of it internally, and ligamentous bands attach to the upper margin, and inner surface, which is concave, whilst the outer is everywhere convex; the convoluted form of the whole cartilage on either side includes one-half of the elastic fibrous frog; whilst continuing to be reflected backwards, the upper cartilaginous margin terminates by giving origin to the tendinous bands forming the two ridges of the frog, one on either side of the cleft, immediately anterior to which these bands coalesce and form what is found to be the denser part of the frog, and are seen piercing downwards and forward, to be inserted into the plantar fascia and the anterior concavity of the coffin bone ; the position is represented externally by the point of the frog.

We have here traced the cartilages on either side, as they recede from the lateral anterior margin of the coffin bone, attaching to the lower condyles of the coronary bone, protecting the pedal joint, and then continue, the inner framework serving instead of bone, which is discontinued, where the axis of bearing is passed ; thus the cartilages, with the hoof, give the form to the foot posteriorly, as the pedal bone fills up the space anteriorly, and by bending round, they become connected with the bands of the frog, which are attached to the centre of the pedal bone inferiorly, forming a double bow. We have passed unnoticed the elastic fibrous bands, which, with interlaying fat-cells together, make up the bulk of the frog, and form the bulbs of the heels. The two lateral halves of the frog, posteriorly, form a double elastic cushion, and are slenderly connected in their posterior centre, represented by the cleft, and are also loosely attached by fibrous bands above, whilst anteriorly the tendinous structures taper, become condensed, and send off attaching bands laterally to the semilunar crest, and a still more considerable one on each side in connection with a slip of the expanded perforans tendon, which are strongly inserted together, into the inner surface of the basilar process, and the pedal cartilage.

Persuaded that no mechanic whose understanding is limited to the handicraft work alone, will ever succeed well in the practice of shoeing horses, and yet, believing from the extent and varied kinds of knowledge required, that few will be able to master more than a special part, and knowing that, probably for years to come, not one horse-shoer in fifty will have the opportunity to learn anything about the foot, (not through want of capacity on the part of farriers to learn, but for reasons, and through the operation of causes not easily defined, but still more difficult to remove), it is through thus seeing facts the more our aim to adopt phrases and modes of exposition such as will be generally intelligible, and still anticipating that which Bracy Clark prognosticated fifty years ago, "when the art has made some advance, improvements will go on faster," the aim now must be to adapt present means of diffusing knowledge on the most urgently required details.

When, however, the most has been done that can be accomplished, to make the veterinary student proficient in the practice of shoeing, and the working farrier intelligent on the structure and functions of the foot, the number of men who may be expected to reach proficiency in the whole, will not be great, the few nevertheless, will be like the little leaven that leavens the whole lump, and in this respect veterinary science and art is not exceptional: the law prevails in all human knowledge. It is very desirable, also, that owners of horses, and men who have charge of them, should possess more correct knowledge on the proper way to manage feet than prevails now, and, as a matter of course, in relative degree as right knowledge becomes established amongst veterinary surgeons as a body, so it will show itself in a more popular form.

It is essentially the provision of veterinary science to enlarge the field of right knowledge on this most important branch in its allotted sphere. And let us try to exclude error, as the only way to advance truth, thence will spring correct ideas, which will radiate. All that is laudable and profitable to the public generally, and to the small body of veterinarians specially, will spring from the banishment of ignorance and prejudice.

In no age in the history of veterinary science have we evidence of such conflicting opinions, and such an unsettled state of knowledge on the economy of the horse's foot and the art of shoeing, as has prevailed in England during the first fifty-five years of the present century. That which was nobody's business, has become every one's province to possess, so that everybody thinks he knows more than others, talks and writes, whilst lame and worthless horses are being multiplied; we may express, as our conviction, however regretable the fact, that this branch of the veterinary art has suffered more within the present century than can reasonably be expected to be redeemed during the remaining part of it.

As this subject has been treated in the different numbers of the *Edinburgh Veterinary Review*, during the past five years, repetition of what has been there produced will not, to any considerable extent, be had recourse to; those who read the *Review* will do well to refer to parts in which the physiology of the foot is described; and in the

meantime we will endeavour to advance on our course by
new methods of analysis, and exposition of views.

It is found to be true, however paradoxical to the mind,
that the art of horse-shoeing, and the prevention and cure of
foot diseases, are subjects so allied that no attempt to separate
them has effectually done so, because, from the nature of the
subjects, the whole must be taken under cognisance together.
Physiology and pathology, or the science of healthy action.
and a knowledge of diseased states, necessitate one unbroken
train of inquiry. Foot disease and lameness, which destroys
a larger proportion of horses than any disease to which the
animal is subject, besides impairing the working powers of
the larger proportion of those which are kept in work, must
be investigated along with prevailing customs of shoeing,
when it will be found that the one stands related to the other
as cause and effect, and the sequence of these will be according to that of the agencies; for instance false modes of shoeing, which, interfering with functions, produce disease; whilst
rational application of art may be made the means of restor-
ing functions, when an all-bountiful nature begins to restore
the mischief done; hence shoeing is a common cause of temporary pain, and it causes permanent derangement of the foot
if the system be not duly amended; whilst the art scientifi-
cally applied is capable of conserving horses' feet in their
normal state, with some occasional and accidental exceptions,
and. when good execution is made to follow bad practice,
shoeing becomes the restorative means; thus it is a surgical
appliance, in the case.

The difficulty experienced in trying to instruct non-pro-
fessional readers by reference to anatomical details, is perhaps
more imaginary than real, though anatomy and physiology
cannot be thoroughly gone into locally, and those who advance
the knowledge of an organ, or a region, do so usually by con-

centrating their inquiries on the part specially, after the whole has received the ordinary share of attention. The same argument, however, applies to all knowledge, and if nobody ventured to learn any abstract truths, because opportunity was wanting to go into the whole length and depth of the matter, in such a case the world would be in a miserable state of ignorance. Now the difference between the few scientific men and the many of the world in the knowledge they possess, is one of degree only, and the more exact and extended the knowledge of the latter is, so relatively will these push their pioneers onwards, and furnish recruits to their ranks.

Having, in recently published papers already referred to, treated on the foot of the horse functionally and physiologically, in which way it must be studied and regarded from the knee and hock inclusive, in fore and hind limbs respectively; and having entered at length on some of the most important details of these phenomena, I shall not trespass on the reader's time by inviting him to follow me over the same ground again in this place, but shall limit my observations to the foot as it is popularly recognised by practical horsemen and horse-shoers.

It has been already said that that complex organ, the foot, is endowed, like every part of the animal frame, with its bony structure, and furthermore with an outer supporting one composed of horn, called the hoof. Thus limiting the sphere of inquiry, we cannot lightly pass over these structures and their relations.

On the right understanding of the economy of this twofold solid construction, greatly depends the capacity for appreciating the functions of the foot.

Other phenomena, of the kind noticed, make up the marvellous combination of pliable structures which are found

running in such variously assigned directions, as to confer
the greatest possible strength, with the most perfectly
adapted resiliency to the whole foot.

The coffin bone is the broad arched structure with which
all the yielding parts are connected, and with which all their
functions are blended. We may refer to the fact, that the
coffin bone is the most important of any one in the skeleton
of the horse for the veterinarian to become acquainted with
thoroughly, owing to its elaborate construction, and owing to
the very important functions with which it is endowed, and
through its being more than any other exposed to injuries ;·
yet, strange to say, just as complications prevail, so relatively
is the bone little understood.

Much that is bad in the customary practice of shoeing,
and equally irrational in the treatment of horses' feet in
every way, is due to wrong notions being entertained on the
form. of the coffin bone, and respecting its most obvious individual and relative functions. More than fifty years ago
the description given of the coffin bone by our authorities, was that of a deformed and diseased bone, instead of
one of normal condition, and it is very probable, for we
have some evidence in the affirmative to show, that one
single bone accidentally falling into the hands of a clever
writer led to widespread erroneous notions regarding it,
which have gone uncorrected ; hence, become and continued
to be a source of growing misunderstanding on the character
of the key structure of the whole foot.

The coffin bone is commonly described as being relieved
from the plain line at its front and posterior extremities;
nothing, however, is further from the truth than such
description, as it is only when the bone has become absorbed
in parts, through pressure, which, unfortunately for owners
and suffering horses, is a most common occurrence, that the

bone assumes the deformity which has been attributed to it as its natural condition.

" In the perfectly natural foot, the *retrossa* are relieved, or raised a little above the general bearing surface of the bone, by which they have a secondary pressure. . . . If we place the perfectly natural coffin bone upon a level flat board, or table, it will be observed to bear primarily on the quarter, and the inside quarter will take a more decided bearing than the outer. . . . The pince, or front of the bone, will also be found to take hardly any sensible bearing, being slightly turned up, and away from the table, obviously in order that it might more conveniently make the rotation which the foot performs on leaving the ground."—(BRACY CLARK, pages 136–7.)

The foregoing is a faithful description of a coffin bone far advanced in deformity, amounting to disease; such as may be found every day, where old horses are destroyed; the bone which was selected was capable of being turned to good account, only as a pathological specimen.

It is less surprising to find an enthusiastic inquirer, in 1809, drawing wrong inferences from first impressions, on seeing a solitary phenomenon, than that his rivals and critics should not, from their number, when excited to move, have discovered his errors; but, instead, some attacked the good, and the less correct results of Clark's labours alike, others believed all that he imagined, but none took the right course,—that of separating from the philosopher's produce, the grain from the chaff.

On the false assumption that the form and functions of the coffin bone are such as has been referred to, indefinite hypotheses have arisen as to how the foot should be shod, and shoes of curious forms have been applied in accordance with the views entertained by the authors, amongst whom we have had, in number constantly multiplying during the last fifty years, not only veterinarians and amateurs, but ironmongers, and men possessing all sorts of knowledge but such as belonged to the subject, have been alike persistent in pressing their claims for patronage, in favour of some peculiarly formed piece of iron, and their views on shoeing in general.

That anatomists should not have been able to establish a common accord on the normal features of the coffin bone is strange, and that physiologists should, by placing that bone on a plain surface, try to find out its bearing points, is to our-selves a mystery. The coffin bone has no bearing surfaces; we know of no bone in any animal that has; as adapted to repose on the ground; it is destined for totally different functions, invested, as it is with other structures, and composed of processes, angles, depressing extremities, and margins; by which means the strongest possible hold is given to cartilages, ligaments, and all connecting and attaching tissues; and it would be as logical to look at the skeleton of the horse, and try to discover that the vertebral spine is adapted to bear the pressure of a saddle and the weight of the rider, as to try to learn how the coffin bone can receive and transmit weight otherwise than through its natural means of connection.

Bones are designed for different functions to that which adaptation of their surfaces to bear external pressure would imply. We will reproduce here some observations and illustrations from the *Edinburgh Veterinary Review* for 1863, to show the connection and relative functions of the coffin bone and the hoof. A large share of the functions of the foot is due to the hoof; anteriorly it is such that it does nothing less than sustain the whole weight and force which, through the limb, are exerted on it. As we have shown, in several published papers, the hoof does not bulge, widen, or expand, when exerted on the ground, but the opposite action is going on; the horse's foot is most relaxed or dilated at the instant when the limb is arriving at its full extension, then it is that the more passive structures are being brought to act, and though, as we have shown, there is no such motion going on as was supposed, we can demonstrate the existence of other movements, better defined, and more constantly going on than those supposed to prevail. By the action which follows the alighting of the foot on the ground, pliable structures are drawn upwards while pressure is being exerted downwards as the several bones revolve, and this follows upon the resilient function which results when the foot is implanted, and which goes on through the whole structure during its instant action.

The annexed illustrations represent the arches of the pedal bone, and the sole, of a fore and hind foot. The drawings were executed from transverse sections through the pedal bones, so as to show the arch most distinctly; the section of the whole hoof was made in the same way, transversely from the upper margin of the wall, down throughout its extent, including sole and frog below.

Figs. 222 and 223 show the pedal bone and hoof of a fore foot.

Figs. 224 and 225 show the same parts of the hind foot.

On examining these structures, the somewhat different functions of fore and hind feet become more apparent to us; corresponding with the external character, and what may be observed in the action of the horse. The fore foot sustains most weight, covers a larger surface than the hind, and the arch, both of sole and bone, is scarcely so high in the fore as in the hind foot, whilst its breadth and sustaining power is greater. In the hind, again, as we endeavour to interpret its action, we are aided by observing the structures, which we find corresponding to the more obtuse exterior point, a more considerable concavity of sole and relatively arched form of pedal bone, and that the hind foot is narrower, and more concave—all properties adapting it to embrace a firm hold on the ground.

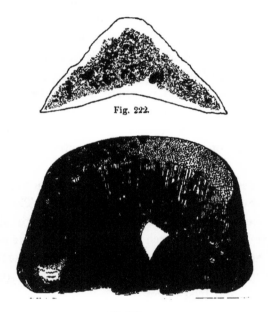

Fig. 222.

Fig. 223.

Fixed rules, to be laid down for preparing the foot, we have none; and it forms a characteristic feature in all the divisions of labour incident to this art, that we require to work by knowledge and observation, and not by measurement.

If it were possible to reduce the practice of shoeing to exact rule, it would no longer be the difficult art it is, requiring knowledge of various kinds, scientific and mechanical, in order to make up the required system.

In concluding our observations on the necessary steps to be taken in preparing horses' feet for shoeing, or adjusting them when going without shoes, we may state that, to preserve proportionate depth of wall, is about the most impor-

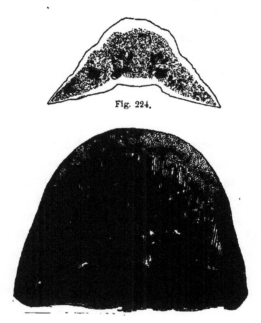

Fig. 224.

Fig. 225.

tant part, when taken alone, to be attended to and well understood of any in the whole process.

When thoroughly understood, the adjustment of the wall to its duly proportioned depth implies more than at first thought appears; it regulates the geometrical figure of the foot, viz., the due degree of obliquity, breadth, length, and depth of heel, and also confers efficiency to its supporting arches.

It should be understood that the inner capacity of the hoof

remains the same, so long as all the organs are maintained in health, it is not affected by the growth of the hoof, all excess of which becomes a store against excessive exertion and consumption, which at irregular intervals is occurring to the horse in his natural state.

The plantar or ground surface of the hoof is endowed with great strength and properties the most effective to resist wear, that is, when horses' feet are preserved and managed as Xenophon directed they should be, viz., kept dry, and accustomed to bear on the hard stones, the texture of the hoof is tough and elastic, yet sufficiently resistant, and its arches are adapted to support their burden.

The large proportion of the hoof which is situated below the organized structure has most important functions to perform, and of course its perfection in form and condition is of importance relative to that of its important offices. If this part be imperfectly developed, as is invariably the case, under bad management in breeding and keeping horses, or if it be reduced by knife and rasp, no amount of iron work can stand in its stead; and, on the other hand, if the hoof is grown to excess, or unequally, and out of proportion, the horse's action, speed, and power become directly affected, even before obvious lameness may necessarily be the result.

The first step in the practice of shoeing to be attended to, is the preparation of the foot. Before removing any part of the hoof, it should be known that the workman is master of his art, and sees at a glance whether any is required to be removed, and, if so, from what part of the hoof. It often happens that a horse comes to us with his foot denuded of its horn, so that there is too little altogether; and yet it may have been so reduced in parts, as to render it necessary that we should give some little adjusting touches; and by removing inequalities, we are often able to give a more extensive and better-balanced bearing surface than previously existed. Thus the word 'adjusting,' or preparing the foot for the shoe, is more applicable than that commonly used, of 'paring out the feet;' indeed, this latter term is thoroughly objectionable, because we never, in the sense implied, pare out the feet at all, the parts detached are not pared away by us, as has been the prevailing custom, and we hold the integrity of the hoof in all its parts to be essential to the foot's health, and to a due performance of its functions; the wall is the chief part we have to act on in the process, which, when defended by an iron shoe, and its consumption is thereby prevented, grows long and deep, and requires reduction to its due proportion at each shoeing, or about every thirty days, even if the horse does not work sufficiently to wear out his shoes in that time.

When it is understood how the feet repose on the ground, and sustain the action exerted on them, the necessity for adapting the shoe in conformity will be apparent. In preparing a foot of which the substance of hoof abounds, there are three points to which attention should be drawn—viz., across the toe, or point, and the posterior extremity on either side. When a horse's foot is much grown up, it becomes what is called long at the toe, and it is at that part more than any other that the instrument is applied for its reduction. At the same time, due proportion is to be given, and the just obliquity of the foot restored or maintained, by adapting the two posterior extremities of the wall to their required depth. Taking the three points above named in order, and for our guide, we carry the rasp over the intervening spaces laterally, which requires to be done nicely—that is, with a clear understanding of requirement and a light touch. One amongst the many prevailing errors in practice consists in rasping and cutting down across the quarters,

making the bearing surface of the foot hollow or flat, whereas
it should possess a fulness appearing somewhat round, as the
eye glances down from heel to toe when the foot is held in the
hand, by which depth is preserved across the centre or quarters, and this is in comformity with the foot's natural structure, and with the normal wear of the hoof when unshod,
and it is, moreover, found successful in practice.

In the preparation of the foot we meet with extremes
in every conceivable form; sometimes the hoof is scarce
through previous reduction, at others excessive accumulation
of horn is found; and though the sole and frog exfoliate when
the foot is in its natural and unshod state, when duly exercised on the ground, and also when properly shod, yet, if we
reverse these conditions, the hoof increases in length and
depth, and the sole and frog become imprisoned, as the wall
grows and tilts over upon itself; in these cases, as a matter
of course, the light touches we prescribe will not suffice;
still the same rule serves to guide. The depth of the wall
being adjusted, unduly thickened sole will be set free, as will
also any morbid growth of the horny frog; and these detaching superfluous parts, being incumbrances, must necessarily
be removed. The continental workman finds over us an
advantage in preparing the foot, by using his buttress—an
instrument discarded in England when the drawing-knife
was made to replace it. There is an important rule to be
observed in preparing the foot, which consists, instead of reducing its depth to excess, as in the ordinary way happens, the
foot should be taken forwards on the knee, and the edge filed
round the toe; this, however, with some exceptions where
there is excess of hoof, will only bear to be done with a fine
rasp or file, nor should there be any deep notch cut in the
toe, as is usually done, to let up the clip, the entire strength
of the wall should be maintained, it being the part which

sustains the greatest exertion of any, and forms the fulcrum in all exertion. The continental shoers do this well by using an instrument which the French call the 'rogne-pied,' with which the outer hard edge of the wall is chipped off, when the buttress does the rest, by paring a broad well-adapted surface for the shoe to be adjusted to.

The 'rogne pied,' or toeing-knife as it was called by English farriers, was formerly in common use with the buttress, by our workmen. Its application, however, was frequently carried to excess, the substance of the hoof being thereby too much reduced; excesses in the same way are not unfrequently committed by the less instructed of continental workmen. The intelligent understanding of the shoer can alone afford sufficient guarantee against abuse, whatever be the instruments used or the rules prescribed.

The opposite to the system we are advocating—viz., of cutting and rasping the hoof flat and hollow across the centre, which is followed by the shoe being made in the same form exposes the coffin bone to be jammed on to the shoe, the point of which being fixed by means of nails, the quarters of the foot form the fulcra on which pressure acts, as the two parallel halves of the shoe posteriorly become two arms of a lever—thus we find the coffin bones pressed into deformity, as will be shown when treating on diseases of the foot.

Besides the conflicting notions entertained as to the way horses' feet should be prepared for shoeing, viz.—how a foot with exuberance of horn is to be reduced to form, or in what this latter consists—the now long-established customs render agreement on method difficult, and necessarily a system will be slow of being brought about; the instruments alone now used by us are ill adapted for the work; the drawing-knife and rasp are apt to rob the feet of their strength and substance, the one by scooping out the concavities, and the other by reducing the outer and prominent parts, but neither nor both together are proper for giving a broad, good bearing surface, or for economising labour in the process. A good continental farrier can get feet ready for shoeing in half the time we can, by using proper tools and forethought, with less physical force.

On the mode of forging shoes, their proportionate substance, and the proper form to be given to them, we shall be brief.

With some few exceptions this part of the work could be reduced to *rule*, difference in size, from such as are suitable for the small, up to the largest horses, constituting the only essential variations; substance and cover is required to be different, according to the employment of horses, their breed and weight, and exceptional formation of feet, taking into consideration also the roads on which they work.

If some plan could be devised by which machinery were made to supply the iron in an advanced stage towards being completed into horse-shoes, a public good would thereby result, and we feel quite certain that water and steam power, under the present state of engineering art, can be brought to do at least one-third of the work which is now done by hand in the making of heavy shoes; and without assuming that anything will ever surpass the best hand-work in the process, we believe that the state of the art of farriery would speedily change for the better, both as it would affect the men employed, and in the execution of the work. It would be possible for much of the heavy labour of forging shoes, which is done for less pay than any similar amount of equally important work, to be kept distinct from the art of shoeing, just as nail making or any of the processes through which iron passes, are separate and distinct departments of industry.

With regard to details on the method of making shoes,

we can admit of some latitude in the diversity, consistently with good execution of the whole process. We profess to have no special shoe, nor do we ever find ourselves in so remote a place from home that we cannot get a man to make shoes to answer our purpose, which we can adjust to horses' feet in all cases, or direct in ordinary requirements how they should be fitted. It is this last part, coming after the preparation of the foot, which determines the good result of the whole operation. There are, however, some rules to be observed for forging shoes, which, unless they be properly made, we cannot succeed well in fitting, and exceedingly well-forged horse-shoes may also be so badly finished and nailed to the feet, that the end is defeated more or less completely.

The fact is, that in the practice of shoeing we are operating on living structures, and though it is difficult for most people to understand this special phase of the art, the truth does force itself occasionally, though, unfortunately, it is the poor horses that really feel where the shoe pinches; losses to owners through shoeing not following instantaneously on the cause, the extent of the evil consequences is not appreciated.

If a man has arms only to wield the hammer, there is not much care taken about whether he has knowledge and a reflective mind, which alone can insure good execution, on which depends the normal action of the horse, with health and ease to his feet.

The rules to be observed in making shoes are—first, to use good iron, which, if of fine grain and tough fibre, will bend under the hammer, even when cold; and can be fullered and stamped when of a red heat, leaving a clean surface, and if the shoes are well forged, they will maintain their form, and afford protection to the foot when they have become thin through wear.

Whilst every man who knows enough of the art of shoeing
to superintend the work, or do it, will appreciate its importance, and the necessity for each part being well done, it is
remarkable that of all the authors who have had their peculiar shoe, not one, to our knowledge, has, in any marked
degree, improved the art of shoeing. Whether we take, for
example, the elder La Fosse, who adopted the short shoe,
or tip, or refer to the several different forms which the late
Professor Coleman adopted and took out patents for, or recall
to memory the names of many men who have also entertained
some special notions on shoes of exceptional form, the good
which the art of shoeing has derived from these is not to be
found ; on the contrary, perpetual confusion is kept up, and
the whole body of working farriers are placed in the position
of an army without a leader. Whilst not a few people
recognise the importance of the art, the majority of those
who keep horses have no right appreciation on the matter,
and the horse-shoer is ordered as if his work was easy to
understand. The art of horse-shoeing is in a more unsettled
state in England than in any other nation in Europe; and
we believe that we are justified in saying that there has
been more want of agreement among men on the subject
during the present century than at any previous epoch in its
history.

There are three requirements of essential importance in
shoes as they are forged; as has been said, they must be
sound and expertly worked, the iron being of good quality,
properly distributed, the nail holes should be of proper form,
rightly placed, and the direction given by the stamp should
be most accurate.

The question as to the form of nail used is also important,
and is intimately connected with the plan of horse-shoe
making ; the apparently essential difference in the form of

nail consists in the variously formed head. The shoes and nails constitute the manufactured material which the artist shoer uses up, and these may be made by expert hands, not necessarily shoers at all; though the workmen fill up their time, when not applying shoes, in making them, still if we could get them equally well made by machinery, it would be a gain in the whole process.

Throughout Great Britain we use two kinds of horse-nails, the old rose-headed pattern, and what is commonly called the countersink; this last kind of nail is of modern introduction, brought into vogue with new ideas on shoeing about sixty or seventy years ago; this nail is entirely confined to the English school; in the whole of Europe, apart from the British Isles, nails of totally different form are used, as we shall hereafter explain, and also by the older Oriental nations.

Writers on the art of farriery have not duly estimated and described how differently formed nails necessarily affect the whole process, and call for modification in the forging of shoes. Those of the old schools are made by the iron being hammered out so that the web presents an almost uniform thickness, which is not more than half that of the outer margin of an English shoe; and they stamp this web of iron with a tool that makes a shallow depression for the nail head. With slight difference, this mode is pursued over the old world, where the art of shoeing no doubt was first applied.

And the practice of shoeing, as carried on by any of the continental nations, will serve to illustrate the views we wish to expose, though the French method, which, above all, merits the name of a well-founded system, as adopted throughout that country, and to a great extent imitated over the world, is that which we shall take most account of, next to the modes in use in our own country.

The French, in forging their shoe, leave the inner margin of the web of their fore shoe (as all people do the hind) as thick as the outer, instead of, like us, working the iron so as to form a thick outer edge, and then to bevel out the inside, making what goes by the name of the seated shoe; our neighbours use a stamp which has a four-sided obtuse point, which forms a perfect countersink, into which the head of their nail fits; the latter being made in a steel dye, exactly corresponding with the nail-hole in the shoe. The French method has advantages which favourably contrast with our custom, inasmuch as the nails can be more accurately applied, from the broad form of the countersink, they have the stronger hold, and less thickness in the web of the shoe is required, and must necessarily be given, by which, and the difference in working the iron of almost uniform thickness, more protection is afforded to the foot, with less weight of iron in the shoe, by from one quarter to a third, than the English shoer uses, and yet, from the mode of adapting the shoe to the foot, the wear is so equal all over, that the lighter shoe will last fully as long as the heavier.

In the process, the first essential difference consists in that of the form of nails used.

We shall not now discuss fully the relative merits of French and English shoes, the object being here to show how incompatible it is to mix systems and adopt parts, or, as some will have it, improve on a plan by imitating an incidental part only.

The comparison between different modes of shoeing will avail more if we go into the description of our own methods, and then refer to the continental systems again.

It is our opinion that the art of shoeing as first adopted in England was of native birth, and not imported; not meaning by this, the origin of shoeing, but that mode of doing the

work, which is so different to that pursued by every other
people, that whilst ideas may have had other origin, and even
shoes may have been brought over, it does not seem that the
English shoer learned this mode of working from any continental or Oriental workman. In the first place, in no other
country do the shoers hold the horse's foot and work at it at
the same time; and though this seems a simple matter, it is
characteristic of the national mode, and to change from one
plan of work to the other, is found to be an ordeal equivalent
to serving a second apprenticeship. An Englishman cannot
do his work by another man holding the horse's foot; nor can
the workmen in neighbouring countries hold the foot, and
prepare it, or nail on the shoe, and finish the work, without
the foot being held. Another part of the process truly
British is that of fullering the shoe; and this was no doubt
at first done for the purpose of marking out by a groove
where the nail-holes were to be stamped; hence the
creased, or, as we now call it, the fullered shoe of these isles.
Nails were also made with the heads to conform to the
crease in the shoe; that gave us the old rose-headed nail
which has never been but partially superseded, and will, we
predict, hold its place in future time as the best English nail
This difference between the Oriental and continental stamping, and the English method of creasing the shoe, led to a
thorough divergence of practice. Workmen are clever men;
they feel their way as they go, and by creasing the shoe, it
would soon be found that the iron must be worked so that a
thick outside margin was left—as we say, a thick outer
edge—or else the fullering would burst, and there would be
insufficient depth for the head of the nail of its peculiar form.
This difference from that of the stamping adopted of
old, brought about another divergence; for, as with the
English stamped shoe, the outer margin must be thick, the

shoe would have been excessively heavy and clumsy, if the same substance had been maintained over the web to the inner margin, and, to obviate this inconvenience, the workman drew away the iron from the inside—thinned the web from the outer to the inner margin.

The custom so far established, became more elaborated. Our smiths were able men at forging iron; their shoes in due time were bevelled out, and hammered up more cleanly; and so far as we have been able to learn, this state of things went on improving, and was not materially disturbed until late in the last century. The establishment of veterinary colleges in France, and the success with which our neighbours had long prosecuted the art of shoeing, led some men of our own country to direct attention there for improvement.

Towards the end of the last century, the first English veterinary college having then been recently established at London, the merits of continental shoeing became more than usually canvassed, comparisons were made between the most approved systems which had long been successfully adopted by our neighbours, and the customary practice which prevailed in the United Kingdom.

The most noticeable difference between all foreign horseshoes and those of British make, consisted in the foreign workmen using the stamp only to make their nail-holes, whilst we in England fullered our shoes, formed a crease, or drew a line for guidance, into which the nail-holes were stamped. Wherever we have found the fullered or creased shoe in use on the Continent, as exceptional to the ordinary method, it was and is still distinguished by its being called the English shoe—" Fer Anglais."

It will be, for time to come, regarded by men who devote themselves to the subject, as a most unfortunate circumstance, that changes were brought about in this kingdom at the epoch alluded to, without sufficient knowledge of the matter. Undue importance was attached to incidental parts in the practice, whilst erroneous notions were entertained and promulgated on what the true character of the French system consisted.

The French method of stamping the nail-holes was allowed by Professor Coleman to be worthy of imitation,

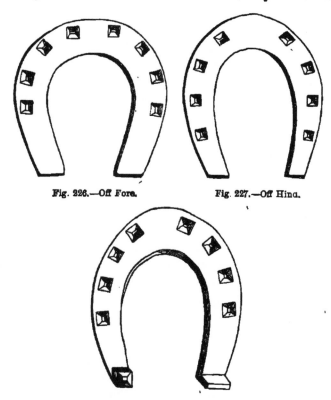

Fig. 226.—Off Fore.   Fig. 227.—Off Hind.

Fig. 228.—Near Hind, with Calkins.

as much preferable to the fine fullering and consequent meagre hold given to the nails by the modes then in use

in England. But Mr Coleman erred in forming an opinion
on the French shoe, when he presumed to enter on a
description of it, without having seen and made himself acquainted with the art of shoeing generally, and of the French
system in particular. Mr Coleman said that the French
stamped their nail-holes coarsely, whereas a more correct
knowledge would have shown that it was eminently a systematic method which they followed, each nail being placed
where most effective, and least in danger of causing injury to
the foot.

On vague notions brought over the Channel, the custom of stamping instead of fullering the shoe was introduced
in England. In his work on the " Foot of the Horse," published in 1798, page 116, Professor Coleman says, "the nail-
hole should be made with a punch, of a wedge-like form, so
as to admit the whole head of the nail into the shoe;" and
" that the head of the nail should be conical, to correspond
with the nail-hole."

The above-described mode of forming the nail and nail-
hole was extemporised by Professor Coleman when he had
everything to learn about shoeing, whether as regarded
English or foreign modes of procedure, as is clearly evidenced
by his book, which was written at the outset of his career,
instead of being deferred until experience had put him in
possession of material knowledge for the work.

Yet such was the position in which the sole teacher of
veterinary medicine in the whole kingdom was placed, that
he had the power, and used it, in so far as the notions represented by his "Principles and Practice of Shoeing" were
accepted, and became diffused, his influence being sufficient
to cause the adoption of all his suggestions in the army;
whilst additionally, or rather primarily, the professorial chair
and the practice pursued at the Veterinary College, were

enough, in a few years, to make Professor Coleman's stamped shoe, with its concave ground surface, and the free application of the drawing-knife to the sole, to become very general throughout the kingdom.

That Professor Coleman was instrumental in carrying out some important changes has been freely conceded, and by none more cordially than ourselves, his insisting on the necessity of a better system of ventilation of stables, and all places where animals were kept, whether temporarily or constantly, did incalculable good; the destroying of glandered horses, and separating cases of infectious disease from healthy animals, with other hygienic measures, were all praiseworthy, though they in no degree affect the question as it regards his modes of shoeing, the treatment of cases of lameness, or his whole course of teaching on the economy of the horse's foot One and all of Coleman's ideas about the foot were an encumbrance to the veterinary student, such as incapacitated him for going calmly and rationally into the matter.

Effects followed causes. The good parts of Mr Coleman's teaching stood the test of time, and now, when it can be shown that the health of our cavalry horses is greatly improved compared to old times, lameness prevails undiminished, so much so, that we never witness a sale of cast-off military horses that is not composed to the extent of about two-thirds of the whole number of lame horses, mostly preventable cases, and many of them curable by the simple application of a better system of shoeing than now prevails in the service.

The army affords the best means for training men up to the highest standard that distinguishes individuals, but in horse-shoeing the reverse has been the case; it has actually afforded a field in which inexperienced men have tried their hands, set at nought whatever was sound of old, and brought their pernicious schemes to bear their fruits.

Before concluding our remarks on the formation of shoes of various kinds, some farther observations on the nails in use, will perhaps help to make the general subject plain, as the form given to them necessitates modification in the construction of shoes.

Three diverse forms of nails representing the main characteristic properties of those in use throughout Europe

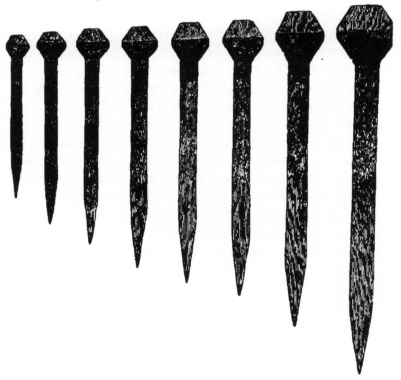

Fig. 229.

require to be noticed, and which the accompanying illustrations are designed to represent. The drawings, executed by Mr A. Brett, who unites with his professional endowments those of a draughtsman of considerable merit, were copied from specimens of the best-made nails of their kinds.

Fig. 229 exhibits an assorted series of French nails obtained from a Paris house, and selected by one of our professional friends of that capital. The eight nails drawn for this illustration include the extreme limits from the smallest to the largest size used in that country, for shoeing very small up to the heavy cart-horse inclusive.

Fig. 230.

Fig. 230 represents a similar series of the old English rose-headed nail, or, as they may not inaptly be called, "the Scotch nails," since it is in the northern division of the kingdom mostly that this pattern of nail has been retained in use to the almost entire exclusion of other forms; whilst in London and over most parts of England and in the army, the wedge-formed nail of Coleman has supplanted the older kind of English nail.

Fig. 231 respresents four nails, *A* and *B* showing these in general use in Scotland, corresponding with the rose-

Fig. 231.

headed nail of old as used in England, and the wedge-formed nail extemporised and introduced into practice by Coleman; these are of the small sizes such as are used for shoeing race-horses. *C* and *D* represent the same diverse kinds of nail of large size, such as are in common use for shoeing the heavy draught horses of Edinburgh and Glasgow, and the dray horses of London.

Fig. 232 .represents one each of the French, the Scotch, and the English wedge-formed nails, of the large size as used for the heavy draught horses. These three nails are

French. Scotch. Wedge form.
1. 2. 3.

Fig. 232.

pointed, ready to be applied, and are drawn to show their flat or broad aspect; and secondly, edgewise; to show the .comparative thickness of the shank, by which combined ex-position the strength of the shank may be in a measure estimated, that depending (the quality of iron and workman-ship being equal) on the broad even form of the shank from the head to near the point, or the part where the clinch

should be formed. A horse-shoe nail should only be of
a thickness sufficient to maintain its position when driven.
If thick at the neck they press more or less in degree and
frequency on the organized structures. We find, by compar-
ing the three nails represented, that the French-made one is
the broadest and thinnest in the shank, is evenly drawn from
head to point, and in these respects the old English nail, or
that which we adopt, and which is in general use in Scotland,
is similarly formed; whilst the modern English counter or
wedge-formed nail is narrower and thicker in the shank,
deviations from the form of the first two described, which
essentially mark the superior fitness of these and the fault ˎ
in the structure of the last, which would do to drive into
wood, where pressure was admissible on all sides.

In submitting these illustrations of three different forms
of nail, all of which are intended for the same purpose, and
are employed in different divisions of our own country and
on the Continent, we do not mean to discuss the relative
state of the art of nail-making in the respective countries
where each differently formed nail is in vogue. All these
models are excellent specimens of workmanship and quality
of iron; it is the difference in systems of shoeing, of which the
form of the nail constitutes a part, that we are reviewing.

The various nails taken as specimens to be drawn from,
were all of the best ordinary samples of their kinds; those
of French make were procured in Paris, as has been stated,
and the rose-headed—the old English nail—now more exclusively used in Scotland than in any other part of the
kingdom, and the Scottish makers being inferior to none in
making that kind of nail, the specimens were taken indiscriminately from amongst those we use; they are made
near Stirling, with one exception, viz., Fig. 231, A, which
was drawn from a small nail as used for race-horses, selected

from those we found in Mr John Scott's forge; and the companion, *B*, is after the spike or wedge form of Coleman, the specimen was obtained at another training establishment, where the narrow, thick tapering nail is preferred to the old sort. The counter-sink or wedge form, from which drawings were taken to compare with the French and Scottish cart-horse nails, were obtained from one of the great London brewery firms, which employs its farriers on the premises, and where the most approved material in vogue is used, and the best workmen are supposed to be employed.

By reference to the figures, it will appear, as examination of the different nails will show, that the French and those in general use throughout Scotland, viz., the old rose-headed nail, bear resemblance, and possess the most essential qualities in common; in breadth of shank they are about equal, the French being somewhat thinner and the more pliable. The essential difference of the two kinds, however, when equally well made, resolves itself into the form of the head, that of the French being solid, filling up a quadrilateral cavity in the shoe, somewhat in the same way as the nail head is let into the tire of a well-made carriage wheel, in which the head of the nail is made to supply the place of the iron removed by the stamp in the case of the shoe, and by the drill in that of the tire. The old English or Scottish nail is made with a flattened head, to adapt it to the crease or fullering, at the same time a fair shoulder is given to these nails by which they are little if at all inferior to the French for holding the shoe, even when the latter is worn thin.

On the counter, or wedge-formed nails, we shall make only a few remarks; as these nails were first extemporised with the English mode of stamping the shoe, and formed a part of that plan; the worth of the nail depends on the merits of the particular shoe, the two forming, as they do parts of a system ; both the shoe and nails are bad together ; the form of the latter, taken by themselves, is bad as compared with the French and the old British form of nail.

In reference to the modes or systems of shoeing, which were adopted of old in our own and neighbouring countries, we will take a passing notice of that particular system, which, with slight deviations, is in use over the greater extent of the old world, and which is generally called Turkish shoeing. That Oriental method is, to say the least, very old, as is evidenced by the simple fact that we have no account of its origin, though, from an unknown period, and still it is applied over many hundred miles from the north-west of Africa to Egypt and Asia; hence we learn of a common practice of shoeing amongst distant empires, kingdoms, and principalities, and have no doubt, from the similarity of method, the history of which we know so little, that it is much older than any of the European systems in vogue which took rise in southern Europe, modified after these Oriental originals.

According to the testimony of travellers, the horses of the desert, and over a great extent of Asia and Africa, go with freedom and ease with the mode of shoeing which the natives adopt, the good result being due, no doubt, to several influential causes, and not to one only.

No European nation has, as far as we know, ever adopted the Oriental method of shoeing, though their armies have occupied large portions of the eastern nations; that fact alone, however, does not signify much, since we have positive knowledge that, as regards the art of horse-shoeing at least, an army may be stationed in a foreign land for years, and adopt nothing of the practice in use amongst the natives, however well adapted to their purpose or the locality. No better proof can be given than that afforded by reference to the English army and its length of service on the Continent at the beginning of this century; the same with reference to the Austrian army in Italy, where we have witnessed the clumsy ill-adapted shoes used on their horses at Milan, Florence, and Naples, the smooth pavement of which cities requires some special provision in the art, instead of which the same customs are adopted there as for horses on the Tyrol and other mountainous districts where snow and ice are common during many months of the year. In truth, horse-shoeing, as regards its importance, soundness, or the reverse of any system in vogue, escapes the attention the subject merits.

Returning to the Oriental shoe, the plan has all the characteristic appearance of an extemporised method for protecting the horse's hoof from wear, prompted by innate wisdom, in adapting means conformably to the economy of the horse's foot, and the surface of ground on which he moves. Still, from trials which we made of that ancient and extensively adopted mode of shoeing many years ago, they did not in the end persuade us that it could be advantageously applied in Europe. It is true, that these trials were mostly confined, in application, to lame horses, and that at a time when our notions on the functions of the foot, and the effects of shoeing, were very different from what they are now.

It was not until recently that I could thoroughly understand how the Oriental system of shoeing is adapted to the regions over which it is exclusively in use; and I am convinced that no method in vogue in Europe could effectually supply the place of the shoe, so admirably suited to the sandy desert. The shoe from which the illustration (fig. 233) is taken, was presented to me about twenty years ago by an English nobleman who brought it from Egypt. After all that one has heard about the African and Arab shoe, and its

relative merits when compared with others, I now see that
the explanation why this shoe is the best that can be devised
there, and not applicable for common use here, is easy.
The thin plate of iron hammered into form, for the most
part when the iron is cold, and which is made to cover

Fig. 233.—Oriental Shoe.

almost the whole of the bottom surface of the foot, is admirably adapted to the movements of the foot; with this the
horse treads on the loose shifting sand, and with the shoe
finds purchase, whilst any of the shoes in use amongst us
would sink into the sand, the horse would slip, and more
exertion would be required, and less speed attained.

I look on it that the Arab horse with the customary shoeing adopted in that country, would have the same advantage
over another horse shod on our method, as horses would have,
in drawing a sledge over the snow-covered plains of Russia,
over others drawing a common wheel carriage.

The French method, as the best system established, is,
we consider, a rational reduction or adaptation of that just
noticed. Our Gallic neighbours of old adopted all that the
Arabs did as regards the adjustment of the shoe to the foot,
whilst the modern open parallel French shoe, which admits a
bearing of the wall of the hoof over its whole circumference,
leaving the concave sole and resilient frog free, is prefer-

able, and most adapted to European soil and artificial roads, yet the mode of stamping the shoe, and the leaving the web as thick inside as out, is traceable from the older type of shoe of eastern origin above noticed.

The English shoe, as far as we can trace its character, has

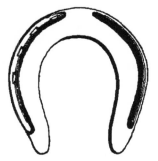 

Fig. 234.—Near Fore.   Fig. 235.—Near Hind.

been distinct in its leading features from any others the history of which we know anything about. Our ancestors seem to have been cautious, and before forming the nail holes, they made a groove around the outer circumference of the iron, to mark the line where the nails were to be placed, at given distances apart, in the groove.   '

It was, no doubt, from the first attempts and by degrees on the part of the English, that the creasing became part of the art of making the shoe; the reason which gave rise to the crease or fullering seems to have been so far lost sight of, that the workman in after-time was esteemed most able, who could make the fullering with the cleanest edge and best form. Fine fullering was in vogue at the end of last century and early part of the present, which led to slanting the stamp and nail inwards, taking slender hold of the hoof, and in order to retain the shoe securely on the foot, the nails were driven high up the wall; that mode was both unsafe and unsubstantial, as also difficult of execution in the application of the shoe.

As has been noticed, our nail-head was flattened conformably to the crease or fullering, sufficient shoulder being given to admit of firm retention of the shoe, when the substance of both it and the head of the nail became reduced by wear.

The early custom of creasing shoes adopted by the English, did not necessarily make very material difference in the principle of shoeing, compared with continental practice. As we have had opportunities of witnessing where, on the Continent, at the present time, the farrier is requested to shoe a horse in the English mode, he understands by that, that he is to make a crease round the margin with a tool, does that, and stamps the nail-holes with his pritchel, instead of with his ordinary stamp, then flattens the head of his usually adopted nails, fits and applies the shoe in his ordinary way, charges extra price, very justly, for the trouble of shoeing English fashion; whilst in all respects the horse is shod according to the ability of the man, in his ordinary way, whether he be a Frenchman, a German, or an Italian. Indeed, in the Austrian army, some years ago, we observed among the officers' horses which were ordered to be shod in the English fashion, that the shoes were fullered, and then stamped with their ordinary quadrilateral stamp, such as we have described as used by the French, they also used their ordinary quadrangular-headed nail—wise practice on the part of the farriers, as it was simply complying with the form, while substantially their own method was carried out unimpaired.

From defects, which distinguished the English mode of shoeing at the latter part of last, and beginning of the present century, which in part consisted in the custom of fullering the shoes on the very margin, the nails took slender hold of the hoof, and from the fact that our neighbours on the Continent never fullered their shoes, but stamped them, and as they were noted for the smallness in the number of their lame horses, as compared with ourselves, the incidental part of stamping was caught up by Coleman, hence the origin of the stamped shoe since then in vogue in England.

Originality, to be practically useful, must result from correct premises, whereas Coleman erred in attributing too much of the fault of shoeing to fine fullering, which admitted of ready correction, without giving up a custom not in the abstract bad, before a better could be established ; and in alleging the chief merit of French shoeing to coarse stamping, these errors led the professor, destitute from first to last as he was of knowledge of kind or degree, to fit him for the work required, to extemporise his coarsely stamped shoe and the wedge-formed nail.

Coleman not having crossed that narrow strait of sea which separates England from France, never saw his mistake in attributing coarseness to the stamping of the shoes, such as are made in the good forges of Paris and Lyons; it could have been easily seen, that from the open space which the four-sided stamp gave, the small punch clearing a passage in the centre of the dye for the nail, a command was given to the workman to pitch his nail in a direct line to where he intended it to pass through the wall and make its exit.

The advantage in this form of nail hole over any other, consists in the facility for driving the nail high or low, taking little or much hold according to requirement; thus the French shoes are stamped with admirable system, the inside nail-holes having a fine, and the outer a coarser position, and with the advantages alluded to, which the shape of the stamp gives, the French farrier adjusts his shoe exactly even with the hoof on the inside, and somewhat fuller on the out, and his method gives the means of taking strong and yet safe hold of the wall in every part.

On the contrary, the spurious custom of stamping, which was forced into practice in England, was inconsistent with good and safe shoeing. The radical faults in the plan are two: firstly, the shape of the nail and nail-hole; secondly, the coarse holds given by stamping too far into the web of the shoe.

The Army Shoe as at present adopted.

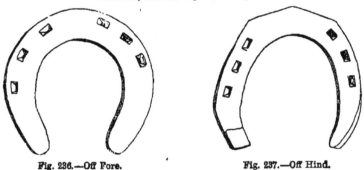

Fig. 236.—Off Fore.   Fig. 237.—Off Hind.

The shape of the nails, as has been noticed, required a thick outer margin to the shoe, in order to bury the long head of the nail, without which precaution it had no hold, but another fault, and as it affected the safety, and became a common cause of lameness to the horses, was the absurd shape given to both nail and nail-hole, which affords no command on the part of the nailer on of the shoe over the direction of the nail. In pointing out faults, the existence of which no man who understands the theory and art of horse-shoeing can deny, we must be understood, as stating what is the tendency, and what the effects of a custom; aware that there are exceptional workmen, who, in making the shoe and holding the stamp, will apply their thoughts, as if they were directing the nail through the hoof, in these cases, care will in a great measure obviate the ill effects; but a custom which requires such care, is full of danger, compared with more simple and rational systems.

There is just one consideration to be given to Coleman's stamped shoe, which gives it a claim in the estimation of workmen before any other, and sympathising as we do for our hardly-wrought and ill-paid farriers, we will give full attention to this part; there is a saving of labour in making this stamped shoe; a saving of many heavy strokes with the sledge-hammer, as compared with the fullering of the shoe, and some saving also as compared with the French method.

The fullering and stamping together of an ordinary-sized carriage horse-shoe requires over rather than under 60 hard strokes with the sledge-hammer, upon the head of a cold tool, whilst to stamp the same shoe with Coleman's wedge-formed stamp, 12 strokes are about the number required; and the French shoe with the obtuse pointed stamp requires about double the last-named number, or 24 strokes. The difference in a day's work of shoe-making, allowing five dozen of shoes for the task, would stand thus:—

Sledge-hammer strokes to the fullered shoes,	3600
The French method of stamping,	1440
The wedge form or English counter-sink stamp,	720

The above figures show that our men are required, in a single day's work at making shoes, to give 2880 more strokes —five times the number—with the heavy hammer in making fullered shoes, than are required for the stamped shoe, and that in addition and apart from the manifold greater. num ber which the whole process of forging the iron requires, and such difference of labour accounts for the silent approval of the otherwise very objectionable mode. Our fullering, as compared with the French stamping, calls for two and a-half times more sledge-hammer blows than their system requires.

In giving a definite opinion as to preference on the special question of fullering or stamping, one is necessarily led to pay some regard to custom, prejudices, and the habits to which workmen in a country have been trained; these points settled, and all things equal, with free choice, we should adopt the French method of stamping for all classes of draught horses, and such as are employed for road work generally; whilst we should use our fullered shoe, as we do with the rose-headed nail, for horses for hunting, training, and, to a great, extent, for riding work generally.

Our motives, in the first case, would be economy of workmen's labour, durability of the work, and conservation of the horse's foot.

Our reasons for using the fullered shoe for the class of horses mentioned, are, that we believe it gives firmer foothold on turf or in the hunting field. Moreover, the fullering spreads the iron, and thereby gives circumference with lightness to the shoe. In neither case, however, do we wish to be arbitrary in pressing our suggestions; as in both ways horses may be thoroughly well shod; on the other hand, the English wedge-formed shoe we never use, and vote its abolition, the economy of labour in making it notwithstanding.

ON THE WEIGHT OF SHOES.

Writers have laid down regulations for the weight of shoes for horses of given classes; we can only, however, comply with that custom to a limited extent. In all cases, the size and weight must vary according to the form and action of horses, and the size of their feet, even amongst those of the same class.

Taking our own practice alone, we could show ample latitude in the weight of shoes, beginning with the race-horse plate of three ounces, up to the cart-horse shoe of large size, such as is used for London dray-horses, and the powerful Clydesdales of Edinburgh and Glasgow, in which cases seventy-two ounces of iron is not an uncommon amount composing a single shoe.

Blood-horses require from nine to fifteen ounces of iron to each shoe. And the general run of mixed-bred horses, including all those used in the army service, will require every grade of weight from eight or ten up to thirty-two ounces.

Agricultural horses, throughout England and Scotland, vary much in their stamp. The latter are generally the stronger, and require the greatest weight of iron to shoe them, though those used in the midland counties of England are probably about equal in weight. Shoes for these various descriptions of English and Scotch farm horses of the strongest breeds will average about 45 ounces each. Whilst in Cleveland, and the eastern and southern counties of England, where a coaching-bred or smaller class of cart-horse is used, the shoes will average a fourth less than where such pure-bred cart-horses are employed.

What has been said on the weight of shoes has been more to show the impossibility of giving any fixed rule on the matter, than to attempt to establish one.

In the process of making shoes, the iron should be drawn level, no uneven thickness to be left at the toe or other part. And if the suggestion already made about the kind of nails be observed, no excess of thickness of the outer margin of the shoe will be required ; neither will there be need for hollowing out the shoe except to a slight extent ; and that, not of necessity for leaving space between the sole of the foot and the shoe, but more for the purpose of lightening the shoe; cover to a moderately proportioned extent being better for the foot than a rigid thick substance of iron. As each separate half of the shoe is made, it is to be fullered and stamped; the outside first, in which we place four nail holes; very coarse holing not advisable, and therefore a clear open fullering, leaving the outer 'edge of the shoe straight, should be aimed at, and in stamping, the tool should be held upright, and not so as to pitch the nail holes inwards. Holes pitching in are dangerous, as in that case the necks of the nails will press more or less against the *quick*. The inside of the shoe being next forged, is to be fullered and stamped in turn, this should be finer fullered than the outside; the same directions hold good regarding the direction to be given the nail holes. We are supposing a saddlehorse's fore shoe to be in hand, in which the nail holes should be distant two inches and a quarter from the outer heel extremity, and two and a half inches from the inner, and about one inch and an eighth may separate each nail hole on the outside, and one inch and two eighths those on the inside, and an inch and five eighths to two inches. According to the size of shoe may be the space between the outer and inner toe holes. These directions for making a plain fore shoe may be generally adopted, size not altering the case, the number of nails we only alter in the case of very small feet, when six nails will suffice; and we sometimes increase the number to eight for large coach horses. It must, however, be remembered, that with the variation in the size of shoe so are the nails varied from Nos. 4 and 5 to Nos. 10 and 12; thus the number of nails required does not differ much in different sized shoes, the difference in size of nail brings the effect equal in the different cases.

What has been said about working the iron level, the fullering and stamping of the fore will serve for the hind shoe.

In other respects, however, the form and substance of the hind shoe require to be different, the iron is kept square in working it, more substance and less cover is required, the shape of the hind shoe must be like the foot, more pointed and of less breadth. The heels, too, should be turned up, and we prefer to turn up both heels in all cases, instead of leaving the inner heel smooth, or, what is more common and more injurious, left thick ; the hind feet, with their strong concave ground aspect and pointed form, act powerfully on the ground in governing the action of the horse.

The custom of applying toe-pieces to the strong draught, as well as cab and some of the carriage horses, is peculiar to Scotland. And, firstly, we will make a few comparative observations between the practice of shoeing the dray-horses in London and that of the heavy cart-horses in Scotland ; the weight of iron used is about the same for the four shoes as they are made in London and Scotland, but the iron is differently placed, and so is the bearing capacity for the foot of the horse to stand on. The London cart-horse's hind shoes are left with a thickness of toe at least double the substance of the quarters; the heels are turned up, and when the horse is shod, he stands on a triangle, the three points of bearing being the toe and both heels. Their fore shoe is also devoid of breadth of toe, and is left thick at the heels, but not turned up.

The Scotch shoe, more to the purpose, is forged level and fullered, the heels turned up, and left strong. The toes of the shoes are flattened down and left pointed, to receive the toe piece, which, for town cart-horse work, is four inches long, cut off a bar three-fourths of an inch to seven-eighths square; these weigh twelve to fourteen ounces each; the toe-piece is welded on when the shoes are fitted.

On fitting the shoes the success of the whole operation

mainly depends. And having dwelt much on the condition of the foot, and how to prepare it, we shall not have to take up much time here, for though this is a nice operation, yet it only admits of being effectively dealt with by the learner, at the forge. After adjusting the heels of the shoe, either by cutting them level or turning them up, clear out the nail holes, take up a clip at the toe, and put each parallel half of the shoe in the exact form of the foot, and approximate the shoe to the required breadth, then letting the left hand regulate the position of the shoe, the hammer falling by force of the right hand, gives the form and surface, adapted to the foot and ground, which, according to the intelligence of the worker, is required. No placing the shoe on the beak of the anvil, to elevate the toe, should be adopted, the shoe being held by the tongs, near the outer heel, the hammer should pass across it behind the clip, the left hand being meanwhile slightly elevated, and then bringing the hammer down, each limb of the shoe, depressing the left hand as the hammer approaches the heels, gives a full bearing surface across the quarter, which should not be uneven, but exactly such as the normal foot maintains when the unshod horse is free. The shoe is to be always full to the foot, in obedience to the law that the pedestal should possess greater capacity than the column it sustains. We have omitted to say anything about hunting shoes; of late all manner of forms have been given to these; whilst we make slight distinction between the mode of shoeing a horse for training, for hunting, or to be ridden in Rotten Row.

CHAPTER XX.

ON THE DISEASES TO WHICH THE FEET OF HORSES ARE SUBJECT.

Contraction of the foot.—Flat or convex soles.—Thrush, its causes and treatment.—Canker.—Corns, their connection with horny tumours.—Sand-crack.—False-quarter.—Fissure.—Keraphylocele.—Seedy toe.—Over-reach.—Treads.—Pricks by nails or other sharp-pointed bodies.—Quittor.—Founder, acute and chronic.—Navicular disease, or navicular joint lameness.

THERE is no department of the veterinary art which calls for more attention at the present time than that concerning the condition of horses' feet ; and, as regards the veterinary surgeon's practice among horses, the art of shoeing has formed, and will form, a most important part of the business of a large proportion of veterinary surgeons practising in large towns ; to all of whom the cases of lameness, their character, causes, and remedies, will, if attended to, take up much time and applied skill.

ON CONTRACTION OF THE FOOT.

This is more of an imaginary than a real state. As we said, when treating on healthy structures and action, there is no alternate expansion and contraction going on in the foot, as has been supposed, during ordinary progression ; neither is there any antagonism kept up by the hoof, tending to constringe the vascular structures, and these, in turn, resisting that tendency.

Complete harmony of parts and counterparts, both in their form and action, subsists between the hoof and all it encircles, the same as between the outer tunic and subjacent structures of every animal in all its parts. To be plain, the question is reduced to this. When a horse's foot, which was once of normal proportion, has become unnaturally small and deformed, how has the change come about, if not by contraction ? Why, by wasting. Like as a man, or any animal which was once fully developed in every part, and when change is brought about by causes, whether from starvation, old age, or otherwise deranged health, wasting ensues and the skin falls in proportionately. Yet, though the general condition of the horse be low, privations, and even old age, exert little appreciable change on the form of the foot of the animal, compared to the more common causes which prevail. It is inconsistent with rational theory, as it certainly is with fact, to entertain the commonly prevailing belief that a clear atmosphere, with the benign influence of the sun's rays, has a tendency to dry up, sear, or otherwise act injuriously on the hoof, which is a living structure, and is supplied from the blood with an abundant secretion for nourishment, and its whole economy. To suppose that a living part should be dried up like as happens with a common board, as has been said (see essay by Professor La Fosse, of Toulouse), is, I submit, giving way to very unphilosophical ideas.

It is the firm belief of the writer, that incalculable harm has been done to our horses, through the entertained groundless notions about contraction of the foot, and again, through imaginary causes inducin that supposed state.

The very strange hypothesis has brought about evil consequences in two ways,—*Firstly*, By treating horses and their feet in a way the reverse of correct; *Secondly*, By being put off their guard, men have neither appreciated nor sought to learn that which is indispensable to the successful management of the animal.

The common agents in the deteriorating process are, excessive moisture to the foot, unduly diminishing the substance of the hoof, and shoes applied without system; and since there is but one right, and an infinity of wrong ways of performing the operation of shoeing, there are many chances to one against this latter part ever being done as required.

Whatever be the relative merits of shoers, the process will be greatly influenced by the general state in which the horse's feet are kept; if firm in texture, and under healthy influences, the power to resist unwelcome pressure will be the greater; when, however, various bad influences are combined, the evil results follow soon and more certainly.

FLAT OR CONVEXED SOLES.

A very similar series of causes, which give rise to the before-described condition, tend also to produce a flattened state of the foot. Breeding, form, action, and weight of the animal, influence the different changes; while the sustaining capabilities of the hoof, and its inherent conservant property of maintaining the symmetrical figure, are dependent on the same hygienic conditions as are required for its first development. All weakening measures tend to alter the physical state of the sole. Softening, paring, and rasping the hoof, want of knowledge, or care, in shoeing, and insufficient exercise, all induce debility, and constitute causes tending to the general bad effect on the whole foot.

By whatever process the sole of the hoof is debilitated, the wall is rendered inefficient for its functions, and either bulges downwards, or collapsing and increase of convexity upwards ensues, accordingly as concomitant influences prevail. Every one who has walked with a soleless shoe, or with the sole soaked by standing in wet until it became

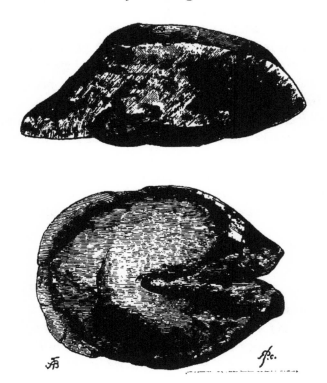

as soft as a sponge, knows that the upper-leather in such cases becomes useless as a means of support to the foot. In the case of the horse, under influences supposed, the pedal bone and all the plantar structures become flattened, as does the hoof, and in conformity with, and by virtue of, natural laws, the most pronounced margins of the bone are removed by absorption, whilst at the same time the energy of locomotive force is proportionately diminished. Having dwelt on the causes, their avoidance will constitute the means of prevention, and an opposite course to such as has been exposed will become the best remedy. A right system of shoeing will be found the best of conservative means, as well as the most effective of curative agents.

Thrush.—This is a diseased condition of the villous membrane covering the fibrous frog; the cleft is the part commonly first affected, and, when neglected, the disease spreads over the whole of the organ to its point and backwards, the horn becoming detached from the bulbs of the heels, and to some extent round the coronet to the quarters. The immediate seat of the disease is the frog; it being a subcuticular affection, and never insinuates beneath the true hoof.

There is no other disease so commonly prevalent in the foot as thrush, and about which there is so much diversity of opinion amongst veterinarians as to the cause; and true it is, which ever view we take, there is always an opposite one, to be entertained and defended almost as strongly, in the belief of any one who has made up his mind differently. Light-formed horses, and even the best bred amongst them, are, under similar conditions, the most commonly affected with thrushes, the hind feet being most liable. Contraction of the hoof has been regarded as the main cause of thrushes. It has been thought that the frog becomes compressed by the narrowing of the hoof. Without further reference to such views, and our grounds of dissent from them, we have only to observe, that this disease is very prevalent amongst horses which are running loose and unshod, and also amongst those where no shoes have ever been applied.

A wet and filthy farm-yard furnishes all those noxious agencies in the greatest abundance which give rise to thrushes, and a general weak state of foot. Reverse these conditions, and keep horses on a firm, well-drained soil, and their hoofs will become almost uniformly excellent physically, as well as firm, elastic, and strong in texture, whilst the arched form of the pedal bone, and the perfection of every fibre in the foot, will be such as to adapt the animal for any work.

Another and common cause of thrush, as it prevails amongst working horses, is justly attributed to impairment of the functions of the foot, through the shoe. When, for instance, the feet, through excess of shortening the toe, and leaving the heels high, are rendered unduly upright, there is a considerable derangement to functions, — one of the common effects being the appearance of thrush.

Treatment.—Remove the causes, and also loose parts of horn from the frog, and put the whole hoof into its normal state, and if the horse is required for work, let him be shod according to rule. Let his stable be dry and clean, with a stone floor; the feet should be washed morning and evening in clean water, and the heels wiped dry at once; and every fourth or fifth day, when the hoofs are dry, a pledget of tow, charged with Barbadoes tar, may be introduced into the clefts of the frogs, and the same pressed into the commissures. If the horse is not required for work, as in the case of young stock, brood mares, &c., let the shoeing be omitted, and all the other things prescribed attended to.

Thrush being, as we have shown, an effect of bad management, in its turn becomes a cause of further derangement in the foot, and therefore no one should consider his

horse in a safe and salutary condition with such an offensive state of the frogs as thrushes present; nor is a horse
in the possession of his full power with that organ in
such a state of ulceration, wasting of the foot being the
usual concomitant of diseased frog. In most cases, when
all the conditions of the foot have been attended to, healthy
action succeeds; in others, such as when horses have been
bred on uncongenial ground, it takes a long time to induce
the normal functions, and especially so, before the tone of
the secreting surfaces can be brought about.

CANKER.

Canker of the foot of the horse is a diseased state of very
peculiar character, in which, primarily, the same local structures are affected as in the case of thrush; and it has by
some been described as another stage, or one arising out of
a neglected and inveterate state of that disease. It is,
however, quite distinct, rapidly extending from the frog
to the sole, and even the laminated structures become involved in the offensive and rapidly destructive ulceration.

The characteristic symptoms of this disease are so strongly
marked, that they can never be mistaken by any one who
has seen a case before. Distinguishable from an ordinary
thrush, by the frog being large, flattened, and spongy to
the feel and appearance, giving off a copious secretion of
most offensively-smelling limpid matter, resembling in kind
and copiousness that which is seen in the worst cases of
grease, to which, in all respects, canker of the foot bears a
resemblance so strong, that the two may be called twin
diseases, often co-existing, and always common to the
same class of low-bred, poor-conditioned horses, such as are
ill cared for, and exposed to wet and filthy stabling. With-

out going so far as to say that any class of horse, under good management, is exempt from this disease, we can most decidedly state that we have never met with a case of canker amongst race-horses, hunters, or in any gentleman's establishment where ordinary average grooming is found ; it is a fact too, that well-bred horses—whatever their class, if of good stamina—are seldom the victims of the diseases, canker and grease, even where the common causes are not altogether excluded.

The treatment for canker consists, in the first place, of the removal of all detaching horn from the affected parts, care being taken, at the same time, not to cut deep, so as to cause effusion of blood, which with the non-experienced operator will be liable to happen, through the deceptive appearances of the sprouting fungous surface, which assumes a horny character, whilst it is endowed with considerable vascularity.

Astringent remedies and mild caustics form the proper dressings. We have found nitric acid and tar the best of all; taking about four ounces of the latter, placing it in an iron laddle, and adding one drachm of the acid, and keeping the mixture briskly stirred with a wooden spatula when heat is evolved, the preparation is then to be spread quickly over the exposed surface, the part being cleansed and dried previously. The surface should then be covered with dry tow and bound up ; or, in cases where the disease is confined to the frog and part of the sole, a shoe may be tacked on, by which means the dressing can be the better retained, by splinters of wood being placed over the tow.

Different practitioners have their favourite and special dressings for the cure of canker; and good results are attributed to the use of a variety of agents, whilst failure not unfrequently attends all our best efforts and means ; in such cases, for instance, as when two feet are very badly affected; or, as we have seen, when all four feet of the horse are equally diseased at the same time, and when it has been impossible to induce him to stand upon the opposite or parallel foot, whilst the one is lifted for the necessary time to be dressed. Nor can casting the horse, for the purpose of dressing the feet, often be made effectual, since the process requires to be repeated every other day, and, at best, will take a long time to produce any permanently good result. Moreover, we believe that there is a constitutional taint in those cases, and experience has taught us that few horses in such condition are likely to pay the cost of their treatment.

CORN.

This is another of the diseases of the connecting structures of the foot in which the hoof participates in the effects.

The prevailing and accepted definition of corn is an erroneous one—viz., that of its being a bruise between the posterior extremity of the coffin-bone above and the hoof below, by which, extravasation of blood is said to ensue. It is nothing of the kind, bruising of the sole does happen in cases of flat-footed horses, while their feet are made still flatter by shoeing and bad management; and in such cases it is possible for the sole to bear on the shoe, fix it at different parts, so as to produce injury to the bone and intervening tissues, when pain and rapidly changing complications follow. In the case assumed above, however, we have not the production of that which has received the name of corn. Corns occur to horses with the best of feet, the common cause being the worst of shoeing. The seat of corn is in the laminated structures at the angles of in-

flection, or, as may be better understood, the extreme point
of the heel. They happen in a similar way under. fast
exertion that a blister does to our heel under hard
marches. The ecchymosis which follows the injury, and
which is called the corn, is nothing else than an after effect,
due to gravitation of the blood-stained serum which is
exuded. The corn is a reality, as its name implies; it
consists in a horn tumour, at the angle above indicated.
These tumefactions reach to various proportions, from that
of enlargement and increased density of the common horn
lamina, to their obliteration, and, in the place, an intruding
growth of smooth horn, more dense than that of any part of
the hoof normally.

The writer first published an account of these horn tumours in 1859-60, when some specimens were presented to
the Museum of the Royal College of Veterinary Surgeons
of London. The discovery led to further observation, and a
more accurate understanding of the whole subject by the
author than he had up to that time arrived at.

This effort of nature to fence out and strengthen, as man
mutilates and weakens, offers a warning lesson to those who
cut and destroy the sole of the hoof: we find that the more
it is scooped away, and the external cavity deepened, so,
relatively, does the intrusion increase upwards, the tissues
and cartilage making way by their becoming absorbed.

These baneful conditions protracted, lead on to further
complications, which, indeed, always progress simultaneously when any one source of injury is perpetually in force.
The most common form by which the approaching crisis
manifests itself in inveterate cases is by suppuration.

This last state seldom arises until after the horse has
endured long suffering from corns; it is not usually until
the internal horn tumours are formed that sloughing of the

parts, and quittor, are brought on. And this is important to be understood, because a prevailing vicious practice is in vogue, under the pretence of exploring, by cutting away the sole of every lame horse in search for matter, and, as is supposed, to give it vent. Blood only, in the case, is found, and with that the searcher is satisfied; whilst mischief is done, such as takes long to repair, even when the patient gets under better care. The proper treatment for corns is a rightly-applied system of shoeing—for which, consult directions on that subject.

SAND-CRACK.

A sand-crack consists of a fissure of more or less extent in length, which always begins at the coronet, in the quarter of the foot over the cartilage. The crack, usually insignificant in appearance at first, is located in the upper thin margin of the wall; the cuticular band giving way, the wall opens in the direction of its fibres downwards. Union never takes place again, but every hour after the lesion is formed confirms and augments the state of the disorder until remedied; inflammation of the skin is set up, the part is painful, and the lips of the wound gape as the tissues swell. The inner quarter is the most common seat of sand-crack, though it happens to both quarters of the same foot in some instances. This ordinary character of sand-crack happens only to the fore feet.

The causes of sand-crack are more varied than those which produce the last disease considered—corn, the strong and good feet often becoming the seat of that lesion and attendant complications.

In this respect the causes of these lesions are common, in so far that sand-crack, like corns, requires the baneful influ-

ence of bad shoeing for its production, with this difference however, that a predisposition almost always prevails
with horses in whose feet sand-cracks appear, and occurring
in those with weak feet,—such as have been bred where
hard or firm sound soil, and liberty to range over it, has
been wanting. English horses generally are, of all domes-
ticated breeds, the least subject to sand-crack of any that
we know of, either European or Oriental; and this observation holds good, though our horses be taken to distant
countries at an early age : the immunity follows on a
perfect development of the whole foot with the growth of
the horse.

Some notion may be formed of the acute pain which the
smallest bursting of the cuticle and hoof at the coronet
gives, by those who have been exposed to causes giving
rise to cracks at the base of the nail; which, however,
it should be remarked, is insignificant, with our hands
moving freely, compared to the forced exertion on the
ground such as the horse's foot undergoes. A small sand-
crack soon acquires larger dimensions, contiguous parts take
on inflammation, and swell ; the wound gapes, and though
no additional splitting of the hoof occurs, the length of the
fissure seems increased by the stretching, and is actually
increased daily by the growth of the hoof, with no pos-
sibility of its reunion. Unrelieved, the case becomes worse,
blood issues under exertion, and, as the wound advances,
with some partially effective attempts to relieve it ; nature,
always provident in fencing out extraneous matter, forms a
secretion of horn in the bottom of the crack, giving rise to
an inner ridge, or, as the French call it a seam ; to get rid
of which, they remove the wall at the quarter, which
is a formidable, tedious, and we believe uncalled-for procedure.

The treatment we adopt consists in placing the part at rest, which, in some cases, calls for giving total rest to the horse for a period of from thirty to fifty days. The most common custom, and one long adopted, is to apply a bar shoe, by which means it is sought to prevent bearing from being imposed on the injured quarter of the foot, which is partially effected by the frog reposing on the shoe, and thus relieving the quarter of some of the burden. The above we have found to be, at best, only a palliation, and not an effectual remedy, and is admissible only with horses used for moderately slow work—for draught.

Of late years, with our more matured experience and system of shoeing, we have been able to give effectual relief by an application to that part of our art, with little or no deviation from our ordinary method or any additional complexity; the feet are, as in other cases, attended to with reference to their general salutary state, and no case has occurred of late where sand-crack has given us any trouble, or which has not healed and the hoof grown down completely sound, commencing from the time we have adopted the proper means.

Giving the horse complete rest, without shoes, when practicable, affords the most simple means of curing him with sand-crack; nothing else being required than treating the feet, as in most cases of lameness we advise. Placing the horse in a loose box, so that he can get free exercise, as has been already repeatedly stated, forms an additional requisite measure for the restoration, and is compatible with the prescribed rest. No binding avails; and all cutting and exploration should be avoided, especially the practice of firing, which is only a source of pain and injury.

False quarter, considered along with sand-crack, may be regarded as an attempted distinction without a difference.

A little confusion of terms is in the case brought into play. In some countries, in Italy for instance, of which the literature on the subject is older than our own, all ordinary sand-cracks are described as quarti-falsi, which implies a condition—viz., when the cracks have become chronic, and the wall exhibits a gaping fissure from the coronet to the bottom, that part of the hoof posteriorly is loose—false, and wants connection in function as well as substance with the front of the foot; and when, from continued or otherwise sustained injury, the secreting structures of the coronet are much destroyed, the absolute breach or great weakness becomes permanent;—hence a permanent sand-crack or false quarter.

FISSURE IN FRONT OF THE HOOF.

A Fissure in front of the foot, though not called, technically, a sand-crack, is a lesion similar in kind; the difference of situation and in the structures affected change the degree, conditions, and character of the lesion under consideration, which commonly happens to the hind feet of draught horses; it is much less frequently met with than that which happens to the quarter of the fore foot, and is also more formidable in its character, and for resistance of curative measures. As in the case of the first described, there is usually a predisposition — weakness, to which may be enumerated the ordinary influences of severe exertion, position given to the foot, and its capacity altered by shoeing, as the constantly recurring causes. The breach in the horn occurs first at the coronet, taking on the same complications and painful condition as has been described in the analogous case; and, from the fact of the horse having to exert his whole power on the solid front of the foot, it is difficult to afford the required ease and relief under exertion, so that treatment, aided with rest, is absolutely called for in bad cases.

As has been said, faulty position given to the limb by means of the shoe is a common cause, it follows that the correction of the vice must be the first step taken in the treatment, and all that has been said on the proper treatment of the foot and shoeing requires to be brought into use in these cases. The sole requires to be left untouched with the knife (as at all times), the wall left deep anteriorly, and the heels moderately lowered—only, however, to a degree to balance with the front, giving the natural obliquity to the foot. Low, firm calkins are proper in the case of hind feet—absence of moisture, and some ointment, or even mutton suet, should be rubbed well into the wall of the hoof every three or four days, to exclude moisture, and check drying of exposed surfaces.

Failing to afford timely relief in this troublesome affection, other complications arise, the horn seam, as mentioned in the case of lateral sand-crack, is in these cases of a more extended form, and a deep substance of horn is found beneath the fissure at the lower part of the hoof, where a corresponding cavity is formed, by absorption, in the anterior convex surface of the coffin-bone. To this state the French have given the name of *keraphylocele*, and for which they operate—viz., remove the whole of the wall at the front, which they do by cutting through it from top to bottom, at given parts, on the inner and outer side of the toe; detaching it from the sole below they tear off the part, leaving the coffin-bone exposed, with only its membranes to support it. This operation is spoken of with approval, which our own observations have not led us to share in. We manage to cure all ordinary, and such as are believed to be curable cases, in a painless and more simple way ;

whilst from the deep-seated ulceration in the bone, the
healing process and the formation of a new hoof to a great
extent must be a process of time; and we also know the
result is uncertain; therefore we merely allude to it, without recommending the formidable operation.   The annexed
wood-cut shows a case of this diseased state in a fore-foot;

the front of the coffin-bone, and the inner cavity of the
hoof, exhibit respectively the deep cavity in the one and the
protruding horn tumour in the other, structure.

SEEDY TOE

Is the English term for a partial separation of the wall
and sole, the vacant space extending more or less up be-

tween the wall and the lower margin of the pedal bone. This condition of the foot is dependent on some previous injury, sustained by the organised structures—to the coffin-bone or tissues covering it, implying destruction more or less of the secreting surface at the part. These conditions may be of a temporary character, or, as is more frequently the case, become, to some extent, permanent deformities; when extensive, there is usually some bulging of the wall at the part affected—this occurs most frequently at the outside toe and quarter. There is commonly a want of the arched form or concavity of sole in connection with these conditions, and there is a corresponding, flatness by absorption, of the coffin-bone. Horses with these defects are not necessarily lame, although it more frequently happens that they are so, and a want of full energy of action must always be a consequent state in the subject so affected; good shoeing and general care of the feet are the means to be adopted.

OVER-REACH.

This consists of a contused wound over the soft bulbs of the heels, usually on one side only, and is caused by the hind foot, whilst in action, over-reaching the parallel fore one; it usually takes place with young horses, especially before they become accustomed to their work in the hunting field. It is only through some false movement that the horse over-reaches. It happens when galloping in deep ground, or when taking a small leap in his stride : it never occurs in the fair gallop; hence the accident is not often met with in race-horses, either in training or whilst running. The injury sustained in over-reaching is not commonly so great as it at first appears to be. The wound.

though extensive, is superficial ; and though the blow has been heavy, from the yielding character of the structures they repel the violence. The cuticle—viz., the soft horn covering—is torn from the true skin and secreting surface, and is detached above, forming a loose pending flap.

Treatment.—In our early days the custom with us was to cut off at once all this detached horn, and then to bind up the wound with tape and tow, wetted with some astringent mixture. In latter years we have adopted a different course, which is found preferable ; it consists in taking off the shoes (always advisable when injury is sustained, which gives rise to swelling in the region of the foot), and, placing the foot in a pail of warm water, continuing to foment it for some minutes, at the same time cleansing the wound by removing all particles of foreign matter ; we then take an ordinary bandage wrung out of warm water, first laying a pledget of tow over the loose flap and whole wound, and press the bandage round the coronet in the usual way, re- peating the same process every 6 to 12 hours. From 48 hours to 3 days is long enough to keep the foot bound up. Bathing it for a few minutes daily, and leaving the wound exposed, is the plan we adopt subsequently, keeping the horse meanwhile in a loose box, and in about six days the bruise passes off, and the dead and detached substance becomes shrivelled and easily distinguishable from sound structures, and may be then cut off, the shoe applied, the hoof rubbed with ointment, and the horse will be fit for exercise or work. It may be asked, why not cut away all detached horn at first ? Because we find economy in retain- ing it ; it forms a better protecting medium than any we can substitute for it ; moreover, it is always well in the case of wounds, where nothing is to be gained by the opposite course, to let all remain until the line between the living and dead becomes well defined : instance, as a comparison, a bad case of broken knee, where we often find our means of perfect cure frustrated through some one having cut off a piece of loose skin, which in the case, would have filled up its former place, or at least, should have been left untouched, and if possible, restored. The same rule applies to over-reach and all similar wounds.

TREADS.

Treads are accidents which commonly happen to horses during the winter, when the heels of their shoes are turned up and sharpened for frosty weather; they occur to draught-horses almost exclusively. The character of a tread is a wound on the coronet, involving a portion of the skin, and usually more or less injury is done to the coronary band. The locality is the inside or front of the foot.

As it is known, any injury to the root of a finger or toe nail is most painful, and of much consideration, when one of similar extent in another part would be little heeded, so with the foot of the horse. Structures of great importance in the economy of the foot, and yet endowed with little vascularity, are wounded in the case; reparation is slow, and the wounds are very painful; death in most cases occurring to the skin at the part wounded.

The horse should be at once put at rest, and the treatment advised in the case of over-reach pursued. There is this difference, however, between the two modes of injury, that in treads deeper seated structures are wounded, and the solid, instead of yielding parts of the foot, sustain the injury.

In addition to warm-water bathing, a linseed-meal poultice may be applied for the first three days in a case of bad treads; after which, and in all slight cases, the bandage, as already recommended, may be relied on. In bad treads we sometimes find that fourteen days elapse before the wound is clean, and all the slough cast off, which, when effected, the cure is soon completed, by continuing cold-water dressing with moderate bandage pressure, until suppuration ceases, when all coverings may be removed. Irritants should be avoided, though some touches with caustic may be called for in the latter stage. The best in the case we find to be crystalised sulphate of copper, which, from its manageable form and slow solubility, can be applied to the precise spot where granulations are loose and sprouting.

WOUNDS OF THE FOOT BY NAILS AND OTHER SHARP-POINTED BODIES.

The most frequently occurring injuries from these causes are inflicted by the nails in shoeing; the cases of lameness from which are more numerous than is generally known, even by shoers themselves.

It is not the palpable accidents of pricking horses that are of the most frequent occurrence, but the lesser and less obvious effects of proximity of the nails to the organised structures; the hoof being at the same time weakened, pain, more or less acute, results, with a variety of effects depending on other conditions. In all these cases of wounds of the foot, the first thing to be done is to remove the foreign body—let all that offends be cleared away, and no exploration, as a rule, is called for, or advisable. Farriers usually do much injury by cutting and exploring, without any notion of what they are in search of; the pretended remedy is commonly an aggravation of the injury; the hoof being destroyed and the foot deeply wounded, whilst no knowledge is gained, and no possible or expected consequences are prevented, but are often brought on by such a course; therefore we content ourselves with the removal of the cause, and in the majority of cases do no more than have regard to the general health of the foot.

The removal of stubs of wood calls for special dexterity ; and the proper after-treatment, when the foreign body has been dislodged, is to apply a linseed-meal poultice, repeating it once in twelve hours, and continuing for three or four days. It sometimes happens, despite all our care, that parts of a stub will be so embedded in the tissues as to evade detection, in which case the suppurative process, which will ensue by the fourth or fifth day under the prescribed measures, will lead to the detection and easy removal of the offending part. When matter is formed, and the wound becomes clear of all extraneous bodies, water-dressings are the best, consisting of pledgets of wet tow laid over the wound, and retained by a bandage.

QUITTOR.

Quittor is perhaps the most painful disease to which the horse is subject. It is another of the disorders that should be prevented; but since it is the nature of our calling to remedy evils that have happened, as well as devise means of prevention, we must prepare to take cases as we find them, and restore the suffering patient in the best possible manner.

As we have, it is believed ; done a good deal in the way of prevention, and establishing a right course of treatment of quittor, so far as our practice and teaching has extended; it may be advisable to go into the details of the disease, which happily is not amongst the most frequent of occurrence ; and when present, none but a qualified practitioner should attempt to interfere in the complicated state which the foot presents.

The accepted definition of quittor, does not meet the case according to our understanding of the matter ; in so far as it is regarded and described as consisting of a discharge of matter from the coronet which has been formed within the cavity of the hoof, caused by some injury, and which so pent up, and finding no means of exit below, burrows its way to the top, sinuses being established for its passage. The above, which gives the common acceptation of what a quittor is supposed to be, is, to say the least, a vague account ; but it is more than that—it is incorrect.

We may have any number of cases of matter escaping from the coronet, without the semblance of quittor, according to our views and experience on the subject. And we think it important to be plain on the matter, because right and wrong notions lead to widely different courses of practice, not only where the existence of quittor is agreed on, but under other and less complicated disordered conditions of the foot.

In cases of temporary injury from the shoe, with exertion in addition, to the extent that inflammation resolves into suppuration, the course to which nature points for the escape of the matter is towards the coronet, and not, as has been supposed, to the bottom of the hoof, where, from its non-vascular nature, it would be locked up. Were this fact known, people would not be so ready to cut away the sole of the hoof to the quick, in order to create a depending opening ; they only find blood, and do incalculable injury, whilst the real tumefaction is shown by a bulging swelling above the hoof, between which and the skin pus, when formed, escapes. And if the horse be properly taken care of by removal of the shoe, fomenting the foot, and applying a poultice or a wet bandage round the coronet, the swelling will subside, and the horse be rendered sound in the course of three or four days' longer rest. Here we have none of the essentials of quittor, though men often make something very like one, which ends in consequences equally bad through excessive meddling. To those who ask how is matter to get free from within the hoof if we don't cut down upon it ? we put another question, Do men suppose that nature left such occurrences unprovided for ? no student of Nature's works will entertain such notion.

The fact appears to us to be, that quittor is a state consequent on a deep-seated lesion of the foot, in which the cartilages, or frequently even the coffin-bone, is affected at its posterior extremity indicated by the seat of the disease. Old standing and progressively increasing corns are the common sources of quittor—recent injuries of the kind only superficially affect the connecting surface of the sensitive foot ; but in process of time horn tumours press into the structures ; when ulceration of bone and cartilage goes on, to remove and make way, mortification more or less of these occur ; hence the necessary separation of parts before a cure can be effected ; and but for proper treatment these cases go on from bad to worse, so that a spontaneous cure rarely occurs. A quittor is a sign of a complication of long-standing disease in the foot, or of some serious lesion, such as the fracture of some point of the coffin-bone, or an ossified cartilage, all of which pathological conditions we have found; so that in our diagnosis of the actual state and probable result, the nature of the case must be fully taken into account, the name *quittor* going little towards helping to solve the question as to the extent and character of the disease.

Symptoms and Treatment.—We are sometimes consulted on the case of a horse very lame, where we find heat and swelling over one heel and quarter of the foot, for which we prescribe fomentations and poultices, and wait the expected suppurative process, endeavouring by these means to modify its extent; and in almost every case we succeed in limiting it to a slight discharge, and immediate healthy termination; in one case out of many, however, the characters which immediately precede the manifestation of quittor appear. In the latter case, instead of a slight discharge of healthy pus, and subsidence of swelling and pain, we find a limpid secretion, with an angry-looking small orifice, and increasing pain—all indicating that there are conditions within that must be changed. More frequently, when called to a case of quittor, we find this latter stage has already passed, and the disease of some days', weeks', and sometimes of months' standing. Commonly we have one sinus externally, in old standing cases sometimes two, at a distance of an inch or so apart, one about the centre of the quarter and the other backward. Our first step in the case is to inject an active agent, which is especially successful in destroying the spurious growths, whilst it very sparingly affects natural, firm structures. We take the following chemical composition : 1 drachm of bichloride of mercury, rubbed down and dissolved in one ounce of rectified spirit, to which we add $\frac{1}{2}$ drachm of liqui plumbi acetetis. With this mixture, by means of a small elastic gum (we find this the best) syringe, we press it into the sinus, and in case there be two, place the finger on one to prevent escape thereat of the liquid, the object being to press and infiltrate the whole tumefied part. We never probe the part, but send the fluid to do the exploring work; for though we have a tolerably large orifice externally, that being the channel into which an infinity of smaller tubes empty themselves, these afford no calibre for the passage of a probe, which when forced, makes one. Within six hours of the application of our treatment, the swelling of the quarter sensibly diminishes, and the horse is to the same extent relieved ; the explanation of which rapid change we submit to be as follows. We have in the case performed something like an amputation ; the agent has destroyed the morbid growth, and has isolated the dead from the living; the medicine, after killing, acts chemically on the matter and diminishes its bulk, so as to take off internal pressure, and the swelling subsides ; the injection should be repeated twice or thrice, at intervals of twelve to twenty-four hours, in order to penetrate to all parts of the diseased structures. Sometimes we elongate the compound after the second application, by adding equal or twice its weight of spirit ; we at this stage wait for the slough separating, which takes from seven to ten days, and watching the case, treat it as the rules of surgery suggest ; and when it goes favourably, as is commonly the case, nothing more is required, besides sufficient rest and good care, until the foot is restored and the horse fit for use.

Connected with the slough, in some rare cases, a piece of bone will come away ; this was called by the old farriers the quittor bone, and consists of a fractured portion of an ossified cartilage or exfoliation of bone, the cause most probably in the case, of the whole mischief.

No horse should be allowed to work under any pretence, whilst suffering under the extremely painful state of quittor, though we often see poor hard-wrought horses in our streets in such condition, with blood and matter streaking down the hoof ; but where words of advice do not avail in these cases, the Act for the Prevention of Cruelty to Animals should be put in force, its tendency being equally conducive to the interest of the owner and humanity to the animal.

FOUNDER, ACUTE AND CHRONIC.

The above terms, and several others, are used at present to express the same condition of the feet of horses and the character of the lameness; the word founder implying, in the case of a horse as in that of a ship, the want of freedom to move.

Fever in the feet is also an old term, used to designate similar cases to the above, neither of which, though somewhat vague, convey the positively erroneous notions of the modern terms, wrongly called scientific, which have been adopted; inflammation of the laminæ, and more recently laminitis, are terms introduced, under the notion that the disease and pain is located in the laminated connecting medium between the wall of the hoof and the organic structures.

We discover a totally different locality to be the original seat of disease, and almost an entirely different train of causes which give rise to it, and, as may be conceived, if our premises be correct, a totally different mode of treatment to be indicated.

The plantar region of the foot being the affected part in all cases which the foregoing terms are meant to indicate, both fore feet of the horse are commonly affected at the same time; sometimes all the feet take on in succession an acute, inflammatory state, arising from being alike exposed to causes, constitutional and mechanical, and the inability of either foot to sustain the exertion destined to be equally distributed over all four. The proximate causes are, exposing the feet to moist soft surfaces, such as on farm-yard manure and wet soil, breeding horses on marshy undrained lands, &c. Bad shoeing and hard work constituting the most common immediate causes, the degree of work which may be borne will depend greatly on the first-named conditions, as a horse may be so bred and treated as not to be able to bear exertion under his own weight.

Treatment of plantar lameness, and the symptomatic fever often accompanying it, consists with us in removal of the shoes in the first instance, adjusting the hoofs, which require to be made level, and not allowed to be high at the heels, the joint of the hoof seldom requires reduction, beyond shortening the spreading thin edge, which is best done with a file, by taking the foot forward on the knee ; the sole should be left untouched with the knife. A mild doze of aloes in ball should be given, and clysters at intervals of two hours for the first day, foment the affected feet with warm water, by placing them one at a time in a pail, or in cases where the horse is not able to stand, flannel bandages wrung out of hot water may be applied with similar effect.

The above emollient treatment locally, and attention to diet, which should be spare and regular in its administration, with the aperient measures prescribed, generally effect a radically good change within three or four days. With continued care, the horse being kept on a clean surface, with his hoofs rubbed with appropriate ointment, and properly shod, he will return to work successfully, relatively in a great measure to his previous normal state.

NAVICULAR DISEASE, OR NAVICULAR JOINT LAMENESS.

Ulceration of the navicular bone, which is found present in many cases of long-standing lameness, has, since it was first brought to notice forty-eight years ago, caused a panic amongst the veterinarians of this country, and through exaggerated notions of the cause, prevalence, and character, the horse-owners have been pecuniary losers, our law courts the scene of conflicting evidence often discreditable to our calling, and the result wrong decisions by juries.

Navicular disease is a state that, by proper management of horses and their feet by shoeing, should be prevented. It is not present in more than one case in ten, if so many, where it is pronounced to exist. Nor is the early, or any stage, so hopeless and unamenable to proper treatment as has been supposed; we have attained success, and been gaining evidence of late years, tending to establish the above propositions. We claim merit for the elimination, as far as our practice and teaching has had influence, of the absolutely destructive measures in vogue,—viz., bleeding, setoning, blistering, firing, &c., and instead, attend to those hygienic means for the restoration of the foot which long experience has enabled us to devise, besides the knowledge derived from all available sources, comprising that of writers of past and present times. Shoeing is our sheet anchor, both as the means of conserving the foot and tending to its restoration, whilst primary and collateral conditions are not to be regarded as insignificant, and these have been mostly referred to in the preceding pages.

GLOSSARY.

ADVENTITIOUS. Unnatural, accidental or acquired.

ALKALI—ALCALI. Term for a substance which has properties the reverse of those of an acid, and with which it combines so as to neutralise its activity and form a salt. It has an acrid urinous taste and caustic quality; it changes vegetable dyes to green, renders oil miscible with water, and is distinguished from an earth by its greater solubility.

ALKALINE. Having the properties of an alkali.

ALKALOID. A vegetable alkali.

ALVINE. Belonging to the belly, stomach, and intestines; applied to the fæces or dung.

AMAUROTIC. Pertaining to that blindness which is produced by paralysis of the nerve of sight.

AMORPHOUS. Without any regular form. Shapeless.

ANASARCA. Watery effusion into the cellular tissue. Dropsy of the limbs, &c.

ANASTOMOSE. To communicate with one another; applied to the connection of blood-vessels and nerves by transverse branches.

ANTENNÆ. Applied to certain articulated filaments inserted in the heads of the Crustacea and Insecta, and appearing to be devoted to a delicate sense of touch.

ANTIPERISTALTIC. The vermicular contraction of the intestinal tube when that takes place in a direction from behind forwards.

ANTIPHLOGISTIC. Against inflammation. Applied to medicines, plans of diet, &c., which counteract inflammation by depressing the vital powers.

ANTISEPTIC. Counteractive of putrefaction.

AORTA. The great artery which arises from the left side of the heart, and gives origin to all other arteries except the pulmonary.

APHONIA. Loss of voice.

APHTHOUS. Having aphthæ or blisters on the skin or mucous membranes.

APNŒA. Absence of respiration.

APPENDICIS EPIPLOICÆ. Masses of fat attached by pedicles to the folds of peritoneum which support the intestines.

ARTHRITIC. Pertaining to joint diseases.

ATAXIC. Showing irregularity in the functions of the body or in the symptoms of a disease.

ATROPHY. A wasting or emaciation with loss of strength; defect of nutrition.

AURICLE. Ear-like appendage.

AUSCULTATION. Attending to the sounds in the different parts of the body, in order to form a judgment of the condition of these parts.

AUTOPSY. Examination after death.

BIFURCATION. A vessel or nerve bifurcates when it divides into two branches.

BUCCAL MEMBRANE. The lining membrane of the mouth.

BUCCINATOR. The muscle of the cheek.

CACHEXIA. A bad condition or habit of body, known by a depraved or vitiated condition of the solids and fluids.

CADAVERIC. Belonging to a carcass.

CANULA. Name of a tubular instrument, introduced by means of a *stilette* or *trochar* into a cavity or tumour, in order that, on removing the *stilette,* any fluid present may be evacuated through it.

CANTHARIDIS. Spanish flies.

CARCINOMATOUS. Cancerous.

CARDIAC. Pertaining to the heart.

CATALYSIS. Name given to a force or power which decomposes a compound body by mere contact.

CATHARTIC. A purgative.

CEREBRAL. Pertaining to the brain.

CEREBRIFORM. Having an appearance like brain matter; applied to a form of cancer.

CERVICAL. Belonging to the neck.

CHOLOCROME. The colouring matter of bile.

CHYLOPOIETIC. Belonging to the stomach and intestines.

CINERITIOUS. Like ashes. Applied to the outer or cortical substance of the brain.

COECUM CAPUT COLI. The point of union of the blind gut with the remainder of the large intestines.

COFFIN-BONE. The last bony segment of the limb, which is enclosed in the hoof.

COMA. A lethargic drowsiness.

COMMISSURE. A jointure. Applied to the corners of the lips, to certain parts in the brain which join one lateral half with the other, and to the hoof, when the frog laterally is connected with the sole.

CONGENITAL. Existing at birth.

CONJUNCTIVA. The mucous membrane of the eyelids, and front of the eye.

CORIACEOUS. Tough, leathery. Applied to leaves.

CORNEA. The clear transparent part on the front of the eye.

CORTICAL. Belonging to the bark of a plant. Applied to the outer layer of the kidney and brain.

CRANIUM. The bony cavity which contains the brain.

CRETACEOUS. Chalky.

CUL-DE-SAC. A blind pouch.

DECUSSATION. Union in the shape of an X or cross. Applied to the crossing of the optic nerves.

DEPILATION. Loss of hair, spontaneously or by art.

DEPURANT. Medicines supposed to be capable of purifying the blood, by removing those constituents which interfere with its purity.

DESQUAMATION. Separation of the scurf-skin in the form of scales.

DEWLAP. The loose hanging skin at the lower part of the ox's neck.

DIAPHORETIC. A medicine which increases the sensible perspiration.

DIAPHRAGM. The muscular partition between the chest and the abdomen.

DIPTEROUS. Having two wings. Applied to insects.

DIVERTICULUM. A blind tube leading out of the course of a longer one.

DORSAL. Belonging to the back.

DUCTUS ARTERIOSUS. A vessel leading from the pulmonary artery to the posterior aorta, and which is obliterated at birth.

DYSCRASIA. A bad habit of body.

ECCHYMOSIS. A livid black or yellow spot, produced by blood effused into the connective tissue.

ECHINOCOCCUS. A bladder worm usually met in the internal organs.

EMBROCATION. A fluid application, to be rubbed on any part of the body.

EMPHYSEMA. A term applied to the presence of air in the areolar tissue, or to diseased enlargement of the ultimate air-cells.

EMPROSTHOTONOS. A variety of tetanus, in which the body is bent forward by the contraction of the muscles.

EMUNCTORY. An organ whose office is to give exit to matters that ought to be excreted.

ENEMATA. Injections. Clysters.

ENTOZOA. Worms that live in the animal body.

ENZOOTIC. Applied to diseases peculiar to a district.

EPIDERMIS. The scurf skin.

EPILEPTIC. Anything relating to epilepsy.

EPIPHYSIS. Part of a bone separated from the shaft in early life, by gristle which afterward changes into bone.

EPITHELIUM. The layer of cells on the surface of mucous and serous membranes.

EXACERBATION. Paroxysm. An increase in the symptoms of a disease.

FAUCES. The gorge. The passage leading from the back part of the mouth to pharynx.

FERRUGINOUS. Chalybeate. Applied to medicines, having iron for their active principle.

FŒTOR. A bad smell.

FORAMEN. An opening ; a hole.

FURCULUM. The merrythought. A fork-like bone in the breast of birds.

GANGLIONIC. Relating to ganglia. From the numerous ganglionic enlargements on the sympathetic nerve it has been called the ganglionic system of nerves.

GLOTTIS. The oblong aperture between the vocal cords of the larynx, and through which the air passes to the lungs.

GRANULATIONS. Small reddish, conical flesh-like shoots, that form on the surface of suppurating sores.

HEMIPLEGIA. Paralysis of one side of the body.

HETEROCHRONOUS. At different times.

HYDATID. A cyst containing a clear liquid. A bladder-worm.

HYGIENIC. Relating to those conditions which tend to preserve health.

HYGROSCOPIC. Belonging to the measurement of the dryness or humidity of the atmosphere.

HYPERTROPHY. Enlargement of a part from increased nutrition.

IDIOPATHIC. A primary disease—one not dependent on any other, or on an injury, is so called.

IDIOSYNCRASY. A peculiarity of constitution in which one individual is affected by an agent, which in numerous others produces no effect.

INCUBATION—HATCHING. The time which elapses between the introduction of a morbid principle into the system, and the onset of the disease.

INFERIOR MAXILLARY BONE. The bone of the lower jaw.

INFUSORIA. An order of vermes of the lowest organisation, and found in putrefying liquids.

INGESTA. Substances introduced into the digestive organs.

INTERMAXILLARY SPACE. The interval between the branches of the lower jaw.

IRIS. A membrane stretched across the anterior chamber of the eye, and which gives it its colour ; it is pierced in its centre by the pupil.

ISOCHRONOUS. Taking place in the same time, or in equal times.

LACHRYMAL. Belonging to the tears. Applied to parts engaged in the secretion and transmission of the tears.

LAMINÆ. Folds or leaves. Applied to the folds in the third stomach of ruminants, and to the horny and sensitive folds by which the hoof-wall is attached to the deeper seated parts.

LITHONTRIPTIC. A remedy supposed to be capable of dissolving stones in the urinary passages.

LUMBAR. Pertaining to the loins.

LYMPHATICS. A system of vessels, engaged in taking up lymph throughout the body.

MALARIA—MIASMA. Noxious emanations from the earth, especially in marshy districts.

MAMMALIA. That class of animals that suckle their young.

MEDIASTINUM. A membranous space between two layers of pleura, extended from the spine to the upper surface of the breast bone.

MEDULLARY. Relating to the marrow, or analogous to marrow.

MELANOSIS. A disease in which tumours are developed, containing a large amount of black pigment.

METASTASIS. The translation (perhaps more properly an extension) of a disease from one part to another.

MONILIFORM. Having the appearance of a string of beads.

MULTIPOLAR. Having many prolongations.

NARES. The openings of the nose, anterior or posterior.

NIDUS. Nest. Seat.

NUCLEOLUS. A simple granule within a nucleus.

NUCLEUS. The centre of a tumour or morbid concretion. A minute cell within a cell.

OMENTUM. Folds of serous membrane passing from one abdominal organ to another are so called.

OPISTHOTONOS. A species of lockjaw in which the body is bent backwards.

OS PEDIS. The principle bone of the foot.

OVARIAN. Belonging to the ovary.

PAPILLÆ. A name given to small conical, vascular eminences on the surface of the true skin or mucous membranes.

PAPULÆ. Small pointed elevations of the scurf skin with inflamed base.

PARAPLEGIA. Paralysis of the anterior or more commonly the posterior half of the body.

PARASITISM. The condition of a parasite or of an organised body, whioh lives on another organised body.

PARENCHYMATOUS. Belonging to the texture of a glandular or other organ.

PATHOGNOMONIC. A characteristic symptom of a disease.

PATHOLOGY. Diseased physiology. That branch of medicine whose object is the knowledge of disease.

PEDUNCLE. A flower stalk. Applied to different prolongations or appendices of the brain. The constricted attachment or neck of a tumour.

PERITONEAL CAVITY. The sac of the peritoneum, or lining serous membrane of the abdomen.

PETECHIÆ. Small purple spots which appear on the skin and mucous membranes in the course of certain maladies. They are attended with great prostration.

PHLEBOTOMIST. One who bleeds from the veins.

PHLEGMON. A circumscribed inflammatory swelling, with increased heat and pain, and tending to suppuration.

PLEURAL CAVITY. The sac of the pleura, or lining membrane of the chest and lungs.

PLEUROTHOTONOS. A variety of tetanus in which the body is

curved laterally by the stronger contraction of the muscles on one side of the body.

POLARITY. That property which disposes the particles of all kinds of matter to move in a regular and determinate manner when affected by other agents.

POLL. The highest point of the head, marked by a transverse bony ridge.

PROGNOSIS. A judgment formed regarding the future progress and termination of any disease.

PROTEAN. Assuming different shapes.

PULMONARY. Belonging to the lungs.

PYRIFORM. Shaped like a pear.

QUARANTINE. The time during which men or animals, coming from a country where any contagious disease exists, are kept from intercourse with the inhabitants of the country.

RECTUM. The last or straight gut.

RENIFORM. Kidney shaped.

SCHNEIDERIAN MEMBRANE. The mucous membrane lining the nose.

SCOLEX. The ascaris lumbricoides—an intestinal worm.

SENSORIUM. The common centre of sensations.

SEPTUM VENTRICULOSUM. The muscular partition between the two ventricles of the heart.

SEROUS MEMBRANES. A class of delicate membranes, which form closed sacs, met with in the chest, abdomen, and spinal canal.

SIBILANT. Making a hissing or whistling sound.

SINAPISM. A poultice of which mustard forms the basis.

SPHINCTER. A name given to several annular muscles, which constrict or close certain natural openings.

SPORADIC. Diseases which occur in isolated cases, and independently of any contagious or epizootic influence.

STASIS. Stagnation, without any morbid condition of the fluids.

STERNUM. The breast bone.

STHENIC. Strong. Action. Applied to inflammation or fever.

STYPTIC. A substance used to check hæmorrhage.

SUB-ACUTE. Applied to maladies which last from seven to twenty days.

SUPPOSITORY. A solid medicinal agent introduced into the rectum.

SUPPURATION. The formation or secretion of pus or matter.

SUTURE. The line of union of two flat bones.

TAXIS. Term for the operation of replacing by the hand, without using instruments, any parts that have left their natural situations, as in the reduction of hernia, &c.

TERATOLOGY. A treatise on monsters.

THERAPEUTIC. Curative. Therapeutics is that part of medicine the object of which is the treatment of disease.

THORAX. The chest.

THYMUS-GLAND. The sweetbread of the butcher. An organ situated in the anterior part of the chest in the mediastinum, and usually absent in adult life.

TRAUMATIC. Relating to a wound.

TREPHINE. An instrument for making openings in flat bones. It consists of a circular saw, centre pin and handle, and removes a circular portion of the bony plate.

TRISMUS. A variety of tetanus affecting only the jaws, and causing their spasmodic closure.

TUBERCULAR. That which relates to tubercles, or that is formed by tubercles. Applied in anatomy to rounded eminences, and in pathological anatomy to a morbid production contained in cysts, or loose in the structure of organs. This matter is at first compact and yellowish ; at times calcareous ; afterwards pultaceous, semifluid, and curdy.

TYMPANUM. The drum of the ear.

TYPHOID. Resembling typhus fever. Often applied to a fever in which the agminated glands of the intestines are in a morbid condition.

UMBILICAL CORD. The navel string.

UMBILICATED. Depressed on its summit.

URETHRA. The channel through which urine is discharged from the bladder.

VARICES. Dilatations of veins.

VENA PORTÆ. A vein which receives the blood from the stomach, intestines, spleen, and pancreas, and breaks up again in the substance of the liver.

VENTRICLE. Literally a little belly. A name given to various small cavities.

VERTEBRA. A segment of the back-bone.

VERTIGO. Giddiness.

VESICULAR. Full of or containing vesicles or cells. Vesicular emphysema is that in which the air-cells are enlarged.

VIBRISSÆ. The hairs that grow at the entrance of the nostrils and other outlets. In cats the whiskers.

VILLI. Small papillary eminences on the surface of mucous membrane, and constituted of blood-vessels, nerves, and absorbents.

VIRUS. A morbid poison. A principle inappreciable to the senses, which is the agent which transmits infectious diseases.

VISCERA. The entrails. Internal organs.

INDEX.

ducens nerve, ii. 436.
omasum or rennet, i. 131.
normal conditions of the pedal
ɔones, ii.
—— vesicular sound, ausculta-
;ion of, i. 518.
scess in the lower jaw, i. 72.
—— of the kidney, ii. 53.
—— of the lung, i. 580.
—— of the turbinated bone, i.
552.
sorption, cutaneous, ii. 77.
use of common salt, in plethora,
i. 327.
ari, finding of, ii. 161.
—— natural history of, ii. 159.
—— time required to destroy by
lifferent reagents, ii. 188.
tions of the spinal cord connected
with the brain, ii. 403.
tive aneurism of the heart, i.
365.
—— capillary hæmorrhage, i. 448.
—— principles used in the treat-
nent of scab, ii., 186.
ite founder, ii.
—— red softening of the cord, ii.
98.
iposus panniculus, ii. 68.
erent nerve fibres, ii. 393.
er pains in ewes, ii. 318.
ə of the horse, to determine by
he teeth, i. 53.
in the veins, i. 430.
— sacs, the, i. 462.
umen, action of the gastric juice
n, i. 215.
— as an antidote in mineral
oisoning, i. 45.
— in food, i. 44.

Albumen in the urine, ii. 26.
—— Mr Percivall on, ii. 41.
Albuminoid substances, chemical
characters of, i. 43.
Albuminuria, ii. 40.
Alkalinity of the blood, i. 34.
—— of urine, ii. 22.
Alveoli, i. 79.
Ambustio, ii. 219.
Ammonia in the blood, i. 39.
Ammoniacal urine, ii. 25.
Amphistoma, ii. 343.
Amphistomum conicum, i. 196.
Amphoric respiration, i. 522.
Amygdalæ, i. 107.
Analyses, comparative, of arterial
and venous blood, i. 492.
—— of renal calculi (Furstenberg),
ii. 58.
—— of sweat, ii. 75.
—— of urine, ii. 19.
—— of vesical calculi (Fursten-
berg), ii. 61.
Anastomoses of nerves, ii. 392.
Anchylosis of temporo-maxillary
joint, i. 68.
Anchylostoma, ii. 342.
Anemone nemorosa, i. 291.
Aneurism, i. 419.
—— of the aorta, i. 420.
Angina, i. 555.
Animal heat, i. 493.
"Anæmic murmurs" i. 355.
Anterior perforated space, ii.
413.
—— pyramids, ii. 411.
Anthrax, ii. 275.
—— cases of, ii. 280.
—— forms of, ii. 286.
—— in the pig, ii. 307.

Anthrax, of the extremities, ii. 302.

Antidotes for metallic irritants, i. 225.

Antivermicular contraction, i. 128.

Antrum Pylori, i. 144.

Aorta, aneurism of the, i. 420.

Apoplexy in pigs, ii. 310.

Arachnida, ii. 210.

Arsenic, as a constituent of animal tissues, i. 39.

Arterial hæmorrhage, i. 428.

Arteries, capillaries and veins, i. 330.

—— diseases of the, i. 499.

Arteritis, i. 409.

Ascarides, i. 258; ii. 340.

Ascaris lumbricoides, i. 258.

—— marginata, i. 260.

—— megalocepala, i. 257.

Ascites, sanguineous, ii. 321.

Ass, louse of, i. 199.

Asthenic, ii. 45.

—— hæmaturia, ii. 47.

Atheroma, i. 425.

Atrophy, of the brain, ii. 492.

—— of the heart, i. 368.

—— of the kidneys, ii. 49.

Auditory nerves, ii. 435.

Auricles of the heart, i. 325.

Auscultation, abnormal sounds in, i. 517.

—— directions for, i 497.

—— of the ox, i. 513.

Azoturia, ii. 32.

Bacteria in splenic apoplexy in sheep, ii. 289.

Bandages, application of, for warmth, ii. 110.

Baths, temperature of, ii. 112.

Benzoic acid in urine, ii. 15.

Biflex caval of the sheep, ii. 71.

Bile, the, i. 238.

—— acids, i. 240.

—— function of, in digestion, i. 242.

—— in urine, ii. 27.

—— test for, i. 241.

—— properties of, i. 289.

Biliary calculi, i. 319.

—— fistula, operation for, i. 239.

Biliverdine, ii. 241.

Birds, respiration in, i. 474.

"Bishoping," i. 63.

Black-quarter, ii. 276.

—— in cattle and sheep, ii. 302

Bladder, description of the, ii. 7.

—— worms, ii. 339.

Blain, ii. 306.

Blood, i. 474.

—— Braxy, ii. 292.

—— Disease in lambs, ii. 326.

—— Extracts from Dr Richardson's essay on the coagulation of, i. 481.

—— Blood in urine, ii. 26.

—— Blood particles, the, i. 476.

—— Blood striking, ii. 286.

—— Bloody flux, i. 287.

Blown, i. 171.

Blue Disease of Pigs, ii. 310.

Blue sickness of the pig, ii. 123.

Bones of the foot, ii. 529.

Bot, Mr Gamgee, seur., on the, i. 199.

—— origin of the term, i. 197.

Botriocephali, ii. 340.

Bowel sickness of sheep, ii. 293.

Brain, the, ii. 409.

—— divisions of the, ii. 410.

—— inflammation of, ii. 480.

—— objects on the base of the, ii. 413.

—— parts of the, producing turning or rolling after an injury on the right side, ii. 429.

—— proper, the, ii. 412.

—— sensibility of the, ii. 428.

Braxy in sheep, ii. 292.

—— mutton, ii. 299.

Bright's disease, ii. 40.

Broken wind, i. 628.

Bronchial sound, auscultation of the, i. 520.

—— Bronchial tubes, the, i. 459.

Bronchi, the, i. 458.

Bronchitis, 571.

Brown Séquard on Epilepsy, ii. 458.

Bruit de Claque, i. 366.

—— de diable" i. 352.

—— de Souffle, i. 396.

Brunner's glands, i. 229.
Buck-tooth, i. 80.
Bullæ, ii. 145.
Bullous eruptions, ii. 145.
Burns and scalds, ii. 219.

Cachexia aquosa, ii. 357.
Calculi urinary, ii. 55.
Calf, louse of, ii. 199.
Cancer of the spine, ii. 506.
Cancerous growths in the heart, i. 384.
Canine madness, ii. 451.
Canker, iv.
Canula.
Capillaries, i, 447.
Capillary hæmorrhage, i. 447.
Carbonate of lime in tissues, i. 35.
—— of potash in food, i. 36.
—— of soda in tissues, i. 36.
Carbonic acid given of in respiration, i. 490.
Carbuncular angina of the pig, ii. ·808.
—— fever, ii. 275.
—— quinsy, i. 561.
Carbuncle, transmitted from the lower animals, ii. 150.
Cardiac polypi, i. 376.
Cardinals, i. 68.
Carditis, i. 385.
Canis, i. 74.
Caries, i. 90.
—— origin of, i. 91.
—— symptoms of, i. 94.
Carious teeth, plugging of, i. 93.
Carnivora, movement of jaws in, i. 49.
—— the salivary glands in, i. 110.
Casein in food, i. 46.
—— action of the gastric juice on, i. 215.
Catalepsy, ii. 464.
Cataract, ii. 376.
Catarrh of the stomach in the dog, i. 225.
Catarrhus sinuum frontalis, i. 542.
Catheter, passing the, ii. 38.
Catheters, varieties of, ii. 37.
Cauda equina, ii. 397.
Caul, i. 236.

Cavernous respiration, i. 522.
Cells of the cuticle, ii. 66.
Cellulose, ii. 28.
Centripetal nerve fibres, ii. 30, 393.
Cerebellum, the, ii. 417.
Cerebral congestion, ii. 474.
—— softening, ii. 491.
—— tumours, ii. 495.
Cerebro-spinal system, ii. 396.
Cerebro-spinal fluid, ii. 399.
Cerebrum, the, ii. 412.
Chapped teats, treatment of, ii. 122.
Cheese, acids in old, i. 47.
Chelidonium, action of, on the urine, ii. 29.
Chemical changes and respiration, i. 489.
—— composition of nervous tissue, ii. 395.
—— constitution of the blood, i. 487.
Chemistry of food, i. 26.
Chest, auscultation of the, i. 508.
Chewing, the act of, i. 141.
—— the rasp, i. 96.'
Chicken-pox of cattle, ii. 256.
Chlorophyl, i. 119.
Chockered, ii. 363.
Chocking, in herbivora, i. 157.
Cholesterine in animals, i. 42.
Cholesterine in bile, i. 241.
Chorea, ii. 465.
Chronic cough, i. 617.
—— hove, i. 175.
—— red mauge, ii. 134.
—— nasal catarrh, i. 542.
Chronicum erythema, ii. 121.
Circinate herpes, ii. 145.
Circulation, rapidity of the, i. 336.
Circulatory apparatus, i. 321.
Ciseau, odontriteur, i. 97.
Classification of lice, ii. 196.
—— of skin diseases, ii. 113.
—— of parasites, ii. 337.
—— of the nerves arising from the brain, ii. 433.
Cleanliness, the value of, ii. 108.
Clipping horses, ii. 95.
Clothing and cleanliness, value of, ii. 108.
Coagulation, changes in the blood in, i. 475.

Coagulation of blood, i. 477.
—— Professor Lister on, i. 482.
Coccyx, i. 233.
Cœcum, invagination of the, i. 277.
Cœcum, i. 30.
Cœnurus cerebralis in cattle and sheep, ii. 350.
Colic, i. 262.
—— causes of, i. 262.
—— results of, i. 270.
Colitis, i. 287.
Collapse of lung, Dr Gairdner on, i. 574.
Colour, i. 230.
—— of hair, ii. 79.
—— of the skin, ii. 67.
Columnar epithelium, i. 209.
Columns of the spinal cord, ii. 399.
Coma, i. 189, ii. 489.
Comatose stomach staggers and sleepy staggers contrasted, i. 190.
Common salt, the quantity of, in different structures of the body, i. 31.
—— the value of, as an article of diet, i. 31.
Complicated diuresis, ii. 32.
Composition of urine, ii. 19.
Compound fracture of the lower jaw, i. 76.
Concentric aneurism of the heart, i. 365.
Condy's disinfecting fluid, i. 30.
Congenital malformations of the heart, i. 362.
Congestion of the lungs, i. 568.
Constipation in dogs, i. 261.
—— in young foals, i. 261.
"Constitution balls," &c., i. 619.
Contagious pleuro-pneumonia of cattle, i. 592.
—— typhoid fever of cattle, ii. 259.
—— typhus, i. 594.
Constituents of the urine, ii. 8.
Contraction, ii. 577.
Convalescence, i. 291.
Convex soles, ii. 579.
Corda tympaui, ii. 439.
Corium, ii. 67.
Corns, ii. 585.
Corpora striati, ii. 413.

Corpora striati, functions of the, ii. 480.
Corpus albicans, ii. 413.
—— callosum, the, ii. 415.
Corrosive sublimate, in attacks of the fly in sheep, 195.
Cortical portion of hair, ii. 79.
—— portion of the kidney, ii. 5.
Coryza, i. 538.
—— gangrenosa, i. 540.
Coughs, auscultation of, i. 528.
Counter irritants, i. 284.
Cow pox, ii. 233.
Cracked heels from improper ventilation of stables, 111.
—— heels of horses, ii. 121.
Cranial exostoses, ii. 493.
Creatinine, ii. 16.
Creatnin, ii. 16.
Crepitant rale, i. 525.
Cretinisms, i. 29.
Crib-biter, cure for a, i. 81.
Crib-biting, i. 81.
"Cripple, the," i. 84.
Crop of birds, i. 145.
Croup, i. 562.
Cruor, i. 488.
Crusta petrosa, i. 50.
Cutaneous absorption, experiments on, ii. 77.
Cuticle, the, ii. 66.
Cutis vera, ii. 67.
Cyanosis, i. 363.
Cynanche malignas carbuncularis, i. 560.
Cystic disease of the kidney, ii. 51.
—— duct. i. 289.
Cysticercus cellulose in the heart of a pig, i. 375.
—— tenuicollis in the pericardium, i. 375.
Cystitis, ii. 54.
Cystocestoid worms, ii. 339.

Dandruff, ii. 66.
"Darn," i. 291.
Deglutition, i. 121.
—— influence of the medulla oblongata on, ii. 424.
—— organs of, i. 121.

Deglutition, the glosso-pharyngeal nerve in, ii. 441.
Delirious staggers, i. 188.
Density of the urine, ii. 8.
Dental papillæ, i. 82.
—— excess of, i. 82.
Dental pulp, diseases of the, i. 92.
Dentinal tubes, i. 91.
—— tumours, operation for, i. 84.
Dentine, i. 50.
—— tumours at the root of the ear, i. 82.
Dentition in the domestic animals, Kreutzer's table of, i. 52.
—— in the ox, i. 57.
—— in the sheep, i. 58.
—— in the dog, i. 59.
—— in the pig, i. 61.
Depletives, i. 267.
Deposits in the membranes of the brain, ii. 494.
Dermis, ii. 67.
Dermoid papilla, i. 84.
Description of the medulla oblongata, ii. 410.
Development of hair, ii. 85.
" Dew Blown," i. 171
Diabetes insipidus, ii. 30.
—— mellitus, ii. 33.
—— produced by pricking the floor of the fourth ventricle, ii. 425.
Diaphragm, spasmodic action of, mistaken for heart disease, i. 354.
Diarrhœa, i. 295.
—— causes of, i. 296.
Diarrhœmia, ii. 325.
Diastole, i. 333.
Digastric muscle, i. 65.
Dilatation of the heart, i. 368.
—— of the stomach, i. 218.
Discharge of blood with the urine, ii. 43.
Disease, the pulse in, i. 344.
Diseases arising from animal poisons of unknown origin, and giving rise to eruptive fevers, ii. 226.
—— arising from special internal causes, ii. 222.

Diseases of, and injuries to the organs of mastication, i. 67.
—— of the arteries, i. 409.
—— of the membrane lining the tooth socket, i. 93.
—— of the nervous system, ii. 449.
—— of the nose producing roaring, i. 620.
—— of the respiratory organs, i. 496.
—— classification of, i. 537.
—— of the reticulum, i. 183.
—— of the skin and hairs, ii. 112.
—— of the teeth, symptoms of, i. 93.
Dislocation of lower jaw in carnivora, treatment for, i. 68.
Distemper in pigs, ii. 810.
Distomæ, ii. 343.
Distoma hepaticum, i. 318.
—— attacks of, ii. 357.
Diuresis, ii. 30.
Divisions of the brain, ii. 410.
—— of the cuticle, ii. 66.
—— of the hoof, ii. 523.
Dochmié, ii. 342.
Dog, louse of the, ii. 199.
—— mange, ii. 171.
—— tick, ii. 210.
Drochmius trigonocephalus in the heart of a dog, i. 375.
Dropping after calving, ii. 313.
Dropsy of the chest, i. 538.
Duct of Nuck, i. 111.
Ductus communis choledocus, i. 239.
Duodenal glands, i. 229.
Duodenum i. 152, 227.
Dust balls, i. 254.
Dysentery, i. 287.
Dyspepsia, i. 219.
—— causes of, i. 220.
—— a result of crib-biting, i. 81.
Dyspeptic palpitation of the heart, i. 352.
Dyspnœa, i. 620, 349.
Dysuria, ii. 34.

Echinococcus vetermorum, i. 319; ii. 349.
—— in the heart, i. 375.
Echinorhyncus, ii. 343.

Echinorhynchus gigas, i. 259.
Ectasis, i. 419.
Ecthyma, ii. 143.
—— of the arms of veterinarians after attending cows in difficult labour, ii. 144.
Eczema, ii. 114.
—— chronicum, ii. 134.
—— epizootica, ii. 265.
—— impetiginodes, ii. 134.
—— Mr Percivall on, ii. 131.
—— rubrum, ii. 133.
—— simplex, ii. 130.
Eczematous or vesicular eruptions, ii. 129.
Efferent nerve fibres, ii. 393.
Elephantiasis, ii. 141, 223.
—— of the ox, ii. 225.
Embolism, i. 409.
Embryo tooth, i. 86.
Emphysema in broken wind, i. 631.
Enamel, i. 50.
Encephalitis, ii. 480.
Encephalon, ii. 409.
Encysted tumour, i. 85.
Endermic application of medicines, ii. 77.
Endocarditis i. 394.
Endocardium, i. 326.
Enema funnels, i. 268.
Enemeta, administration of, i. 268.
English system of shoeing, ii. 552.
Enteritis, i. 281.
—— exudativa, i. 284.
Enterorrhœa, i. 292.
Entozoa affecting the domestic quadrupeds, ii. 338.
Enuresis, ii. 39.
Enzootic and epizootic diseases, difference between, ii. 258.
—— diseases, ii. 272.
—— dysentery, i. 291.
—— hæmaturia of horses, ii. 45.
Ephelis, ii. 220.
Epidemic, the, ii. 265.
Epidermis, ii. 66.
Epiglottis, i. 127.
—— the, i. 454.
Epilepsy, ii. 457.
Epiploic glands, i. 279.

Epistaxis, i. 537.
Epithelium, scaly, i. 8, 107.
—— stratified, i. 125.
Epizootic aphtha, ii. 265.
—— diseases, ii. 257.
Equine lymph. ii. 140.
Equinia, ii. 140.
Eructation, i. 172.
Erysipelas, ii. 123.
—— carbunculosum, ii. 302.
Erythema, ii. 114.
—— ii. 119.
—— treatment of, ii. 122.
Erythematous eruptions, ii. 119.
Esparcet, i. 119.
Ethmoid cells, the, i. 453.
Eustachian tubes and pouch, the, i. 454.
—— valve, i. 326.
Exania, i. 301.
Exanthematous eruptions, ii. 119.
Excessive secretion of urine, ii. 30.
Excitability of the deep parts of the spinal cord, ii. 404.
Exomphalos, i. 305.
Exosmosis and endosmosis, i. 31.
Extractive matters in plants and animals, i. 47.
Extraction of teeth, i. 100.
Extractive matters of the urine, ii. 16.

Face, impetigo of, in calves and lambs.
Facial fistula, operation for diseased tooth leading to, i. 101.
Facial nerve, the, ii. 439.
Fags, ii. 210.
False nostril, i. 452.
Farcy and glanders, ii. 381.
Fardel bound, i. 178.
Fat, origin of, in the body, i. 41.
Fats and oils in the animal tissues, i. 41.
Fatty degeneration of the heart, i. 373.
Fever in the feet, ii.
Fibres of sensation, arrangement of, ii. 407.
Fibrine in animal tissues, i. 45.
——in blood, i. 488.

Fibrinous clots in arteries, i. 411.
Fibro-plastic degeneration of bone, i. 72.
—— degeneration of the jaws, nature and causes of, i. 74.
Filariæ, ii. 341.
Fishes, respiration in, i. 450.
Fistula in ano, i. 300.
—— of the face, i. 92.
—— of the kidney, ii. 53.
—— of the mouth, i. 74.
—— of the reticulum, i. 184.
—— of the stomach, i. 209.
Fitting the shoes, ii. 575.
Flat feet, ii. 579.
Flatulent colic, i. 265.
Flea, the, ii. 201.
Fluke, i. 318, ii. 361.
—— disease, ii. 357.
Fluoride of calcium in tissues, i. 35.
Fly, attacks of the, ii. 194.
Fog sickness, i. 171.
Food, hydrocarbonaceous principles of, i. 39.
Foot-and-mouth disease, ii. 265.
Foramen of Winslow, position of the, i. 237.
—— ovale, i. 326.
—— pervious, i. 363.
Forces which regulate the flow of, blood, i. 335.
Foreign body in the left auricle of a cow, i. 402.
Fracture of the lower jaw, results of, i. 78.
—— of the nasal bones causing roaring, i. 620.
—— of the windpipe, i. 568.
Fractures of the bones of the foot, ii.
—— of the jaws, i. 75.
French method of forging shoes, ii. 552.
Frontal sinuses, the, i. 453.
Frost-bite, ii. 222.
Functions of the cerebellum, ii. 426.
—— of the columns of the cord, ii. 408.
—— of the Corpora geniculata, ii. 429.

Functions of the Corpora quadrigemina, ii. 429.
—— of the hemispheres of the brain, ii. 431.
—— of the medulla oblongata. ii. 420.
—— of the nerve fibres, ii. 391.
—— of the olivary bodies, ii. 424.
—— of the optic thalami, ii. 430.
—— of the pons varolii, ii. 426.
—— of the spinal cord, ii. 402.
Fundus of the stomach, i. 143.
Furuncular eruptions, ii. 149.
Furunculus, ii. 114.

Gad fly, the, ii. 204.
Gallstones, i. 319.
Ganglia on the course of the trifacial nerve, ii. 439.
Ganglionic system, ii. 396.
Gangrene, i. 289.
Gangrenous erysepelas, ii. 123.
Gapes in fowls, i. 637.
Gasserian ganglion, the, ii. 439.
Gastric glands of the stomach, i. 209.
—— juice, i. 209.
—— action of, on medicines, i 217.
—— properties of the, i. 209, 213.
—— acids of the, i. 214.
Gastritis, or inflammation of the stomach, i. 222.
—— in the dog, symptoms of, i. 225.
Gastro-cephalatis, i. 186.
Gastro conjunctivitis, i. 186.
—— enteritis, i. 296.
—— hepatitis, i. 186.
Gastrorrhœa in the dog, i. 225.
Gelatine, action of the gastric juice on, i, 215.
—— in organised structures, i. 48.
Gelatio, ii. 222.
General distribution of the circulation, i. 339.
General symptoms of disease of the heart, i. 348.
Genestade, ii. 46.
Gerlach on the itch and mange, ii. 160.
Gid, ii. 350.
Gizzard, i. 147.
Glanders, diseased teeth mistaken for, i. 95.

Glanders and farcy, ii. 381.
Glands of the skin, ii. 69.
Glandulæ agminatæ, i. 230.
—— solitariæ, i. 229.
Glisson's capsule, i. 315.
Glossanthrax, ii. 306.
Glossitis, i. 21.
Glosso-pharyngeal nerve, ii. 440.
Glucose in the blood, i. 40.
Glycerine in skin diseases, ii. 118.
Glykocholic acid, i. 241.
Goat, louse of the, ii. 199.
Grapes, ii. 141.
Gravel, ii. 55 ; ii. 63.
Grass staggers, i. 178.
—— symptoms of, i. 181.
Grease, ii. 139.
Gullet, inflammation of the, i. 167.
Gum lancet, Gowing's, i. 105.
Gut-tie in cattle, i. 310.
Guttural pouches, accumulation of pus in the, i. 548.
Gypping, i. 63.

Hæmatine, iron in, i. 37.
Hæmatopinus eurystenus, ii. 198.
Hæmaturia, ii. 43.
Hæmione, i. 66.
Hæpatirrhœa, i. 313.
Hair, ii. 78.
——, change in the colour of, ii. 81.
—— concretions in the reticulum, i. 183.
Halsgrind, ii. 138.
Heart, action of the, i. 331.
—— description of the, i. 322.
—— diseases of the, i. 347.
—— foreign bodies injuring the, i. 399.
Heat of different animals, i. 493.
Heaving pains in ewes, ii. 318.
Hemiplegia, ii. 497, 506.
Hemispheres of the brain, functions of the, ii. 431.
Hæmorrhoids, i. 303.
Hepatic duct, ii. 239.
Hepatitis, i. 316.
Hepatization, i. 373.
Herbivora, movement of jaws in, i. 49.
Hernia, i. 304.

Hernia iridis, i. 304.
Herpes circinatus, ii. 147.
—— phlyctonodes, ii. 146.
—— zoster, ii. 145.
Herpetic eruptions, ii. 145.
Hiccough, i. 355.
Hidebound, ii. 69.
Hill hraxy, ii. 292.
Hippobosca equina, ii. 212.
Hippuric acid, ii. 14.
History of shoeing, ii. 520.
—— of outbreaks of smallpox in sheep in this country, ii. 241.
—— of the outbreaks of epizootic aphtha, ii. 266.
Hog cholera, the, ii. 227, 310.
Hogg, the Ettrick Shepherd, on Braxy, 293.
Honey diabetes, ii. 33.
Horn, ii. 513.
—— cells, ii. 514.
—— fibres, ii. 514.
Horripilation, ii. 84.
Horse, auscultation in the, i. 503.
—— bot, i. 196.
—— louse of the, ii. 200.
Hot-air bath, Dr E. Wilson on the, ii. 107.
Hove, i. 171.
—— in sheep, i. 176.
" Humid tetter," ii. 130.
Husk, the, in calves, ii. 366.
Hydatide in the brain, ii. 496.
—— of the liver in animals, ii. 348.
Hydrocephalus, ii. 493.
Hydrochloric acid, action of on bones, i. 33.
Hydrocyanic acid, i. 226.
Hydrogen, i. 28.
Hydrophobia, ii. 451.
Hydro-rachitis, ii. 503.
Hydrothorax, i. 588.
—— recognition of, i. 348.
Hyovertebrotomy, i. 550.
Hyperæmia of the liver, i. 313.
Hypertrophy of the brain, ii. 492.
—— of the heart, i. 365.
Hypoglossal nerve, the, ii. 448.

Icterus, i. 311.
Idiopathic hæmaturia, ii. 45.

Idiopathic peritonitis, i. 286.
Ileo-colic, or Ileo-cœcal valve, i. 233.
Ileus, i. 278.
Ileum, i. 228.
Iliac artery, aneurism of the, i. 422.
Immediate auscultation, i. 497.
Immobility, ii. 489.
Impaction of the crop in birds, i. 195.
—— of the omasum in a flock of sheep, i. 186.
—— of the rumen, i. 176.
—— operation for, i. 177.
—— of the stomach in the horse, i. 186.
—— of the stomach in dogs, i. 193.
—— of the stomach in sheep, i. 185.
—— of the third stomach, i. 178.
Impetiginodes eczema, ii. 134.
Impetiginous eruptions, ii. 135.
Impetigo, ii. 114.
—— colli, ii. 138.
—— erysipelatoides ii. 139.
—— labialis, s. facialis. ii. 136.
—— larvalis, ii. 138.
—— of the face in calves and lambs, ii. 138.
—— rodens, ii. 139.
—— sparsa digitorum, ii. 139.
Incisive foramen, i. 111.
Incissors, deviations in the position of, i. 79.
Incontinence of urine, ii. 39,
Induration of the brain, ii. 491.
Inferior maxillary nerve, ii. 437
Inflammation of the heart, i. 385.
—— of the intestines. i. 281.
—— of the kidneys, ii. 51.
—— of the pleuræ, i. 581.
—— of the substance of the lungs, i. 577.
—— of the urinary bladder, ii. 54.
Influenza, ii. 378.
Infundibulum, the, ii. 413.
Infueoria in the rumen, i. 187.
—— respiration in, i. 450.
Inguinal hernia, i. 305.
Injection of fluids into the nose, i. 543.
Injuries to the arteries, i. 427.

Injuries to the joint between the temporal bone and lower jaw, i. 67.
Inoculation for pleuro-pneumonia, i. 613.
—— for smallpox in sheep, ii. 240.
Inorganic constituents of food, i. 26.
—— salts of the urine, ii. 17.
Inosite, i. 41.
Inperforate anus, i. 299.
Insalivation, i. 66, 107.
Insect powder, Persian, ii. 202.
Insensible perspiration, amount of, ii. 74.
Inspiration, act of, i. 468.
Interdigital canal of the sheep, ii. 71.
Interpeduncular space, the, 413.
Intertrigo erythema, ii. 119.
Intestinal concretions, i. 254.
—— digestion, i. 226.
—— parasites, i. 257.
—— secretion, composition of the, i. 251.
Intestine, definition of, i. 226.
—— strangulation of by pedunculated tumours or hypertrophied appendices epiploicæ. i. 278.
Intestines, length of, i. 227.
—— of ruminants, i. 235.
Intrathoracic choking, i. 167.
Intussusception, i. 275.
—— pathological, anatomy of the lesion, i. 277.
Invagination, i. 276.
Iron in food. i. 37.
—— proportion of in man and animals, i. 38.
Irregularities in the rows of teeth, i. 88.
Irritability of tissue, ii. 386.
Ischuria, ii. 34.
Island of Reil, ii. 412.
Itch in the lower animals, ii. 155.

Jaundice, i. 311.
Jaws, diseases of the, i. 71.
—— movements of in mastication, i. 66,
Jejunum, i. 228.
Joint disease amongst calves and lambs, ii. 328.

Jugular vein, phlebolites in the, i. 440.

Keds, ii. 210.
Kelis, ii. 223.
Kemps, ii. 85.
Keraphylocele, ii. 591.
Kidney, granular degeneration of the, ii. 40.
Kidneys, the, ii. 2.
Kousso, i. 261.

Lactic acid in urine, ii. 16.
Lambs, baneful effects of covering with skins, ii. 180.
—— blood, disease in, ii. 326.
Lamb disease in America, ii. 331.
Lambs, new disease in, ii. 326.
Laminitis, i. 189.
Lampas, i. 11.
Large intestine, i. 230.
—— division of the, i. 30.
—— structure of the, i. 232.
Laryngitis, i. 555.
Laryngo pharyngitis, i. 555.
Larynx, i. 126.
—— auscultation of the, i. 507.
—— cartilages of the, i. 454.
Lateral repeller, Gowing's, i. 100.
Lead poisoning, i. 178.
—— symptoms of, i. 182.
Legumine, i. 46.
Lenticular glands, i. 209.
Lice in animals, ii. 195.
Lichen, ii. 114.
—— albescens, ii. 126.
—— cinereus, ii. 126.
Lichenous eruptions, ii. 126.
Lieberkuhn, crypts of, i. 229.
Lingual glands, i. 108.
Lithiasis, ii. 55.
Lithotomy in the horse, &c., ii. 60.
Lithotrity, ii. 60.
Liver, i. 238.
—— diseases of the, i. 311.
—— form of in the ox, i. 238.
Lobes of the lungs, i. 462.
Local remedies for skin diseases, ii. 116.
Lockjaw, ii. 466.
Lolium temulentum i. 201.

Louping ill, ii. 503.
Louse of ox, ii. 198.
Lower jaw, dislocation of, i. 67.
Lumbricus, i. 257.
Lungs, blood-vessels of the, i. 464.
Lupus, ii. 223.
Lymph, i. 36.
Lymphangitis, ii. 123.
Lymphatic glands, enlargement of in cases of diseased teeth, i. 95.
Lymphatics of the lungs, i. 467.

Mad staggers, i. 188.
Mähnen, ii. 138.
Magenkoller, i. 186.
Maggots in sheep, ii. 195.
Magnesia, carbonate of, in the urine of herbivora, i. 89.
Maladie de Sologne, ii. 321.
Maladie Rouge, ii. 321,
Malignant coryza, i. 540.
Mallenders and sallenders, ii. 130.
Malpighian bodies, ii. 6.
Mange in the lower animals, ii. 155.
—— in the cat, ii. 177.
—— in the dog, ii. 171.
—— in the ox, ii. 170.
—— in the pig, ii. 171.
Manganese in animal structures, i. 38.
Manyplies, action of on the food, i. 142.
Masseters, i. 65.
Mastication, i. 49.
—— process of, i. 65.
—— rate of in herbivora, i. 66.
Maw-worm, i. 260.
Maxillary sinuses, i. 92, 453.
Measles, ii. 226.
—— in cattle, ii. 347.
—— in the pig, ii. 343.
Mechanism of respiration, i. 467.
Meconium, i. 262.
Mediastinum, the, i. 464.
Mediate auscultation, i. 497.
Medulla oblongata, ii. 410.
—— of hair, ii. 80.
Medullary portion of the kidney, ii. 5.
Megrims, i. 432; ii. 474.
Melanosis, i. 121.
—— in the heart, i. 375.

Melœna, i. 291.

Melanotic tumour, causing rupture of the rectum, i. 278.

Meningitis, ii. 480.

Mensuration, application of in chest affections, i. 535.

Mental nervous actions, ii. 388.

Mercurial eczema, ii. 183.

Mercurial ointment in the treatment of scab, ii. 178.

—— ptyalism, i. 118.

Mesenteric hernia, i. 310.

—— veins, i. 40.

Mesentery, i. 274.

Mesocolon, i. 272.

Metamorphoses of the bot. i. 200.

Metastatic abscesses in the liver, i. 317.

Method of using the Roman bath, ii. 101.

Milk fever in cows, ii. 313.

—— in epizootic aptha, ii. 268.

Milzbrand, ii. 275.

Molars, i. 17.

Molar teeth, peculiarities and diseases of the, i. 88.

Monogastric animals, i. 143.

Monostomæ, ii. 343.

Moou blindness, ii. 372.

Morbilli, ii. 226.

Motores oculorum, ii. 435.

Moulten grease, i. 285.

Movement of parasites, injurious effects of the, ii. 336.

Muciparous saliva, i. 120.

Mucous râle, the, i. 524.

Mumps, i. 121.

Musca vomitoria, ii. 194.

Muscles of hair, ii. 82.

Muscular fibres of the heart, arrangement of the, i. 326.

—— tissue, action of the gastric juice on, i. 215.

Mustard, application of to the chest of a horse, i. 571.

Myelitis, ii. 498.

Nails, ii. 560.

Nasal and frontal sinuses, accumulation of pus in, i. 546.

—— catarrh, i. 538.

Nasal chambers, auscultation of the, i. 507.

—— discharge, due to organic disease of the facial bones, i. 554.

—— gleet, i. 543.

—— polypus, i. 95.

—— polypi causing roaring, i. 620.

Natural history of acari, ii. 159.

Navel ill, ii. 326.

Navicular disease, ii.

Neck anthrax of the pig, ii. 309.

—— mange, ii. 138.

Necrosis of the pedal bones, ii.

Needleworm, i. 257.

Nematoda, i. 196.

Nematoid worms, ii. 340.

Nephritis, ii. 52.

—— albuminous, ii. 40.

Nerve-cell, the, ii. 389.

—— fibres, the, 389.

Nerves of the lungs, i. 467.

—— originating from the encephalon or brain, ii. 433.

—— action, ii. 385.

—— system, constitution of the, ii. 389.

—— definition of the, ii. 385.

—— diseases of, ii. 449.

—— tissue, chemical composition of, ii. 395.

Neuritis, ii. 512.

Neuroma, ii. 512.

Nitrate of urea, ii. 11.

Nitrogenous principles of food, i. 42.

Nostrils, i. 450.

Number of pulsations and modifications in the character of the pulse according to species, age, sex, temperament, &c., i. 841.

Number of skin diseases, ii. 112.

"Nutmeg liver," i. 314.

Obesity in heart disease, i. 350.

Objects on the floor of the lateral ventricles, ii. 415.

Odontritia (Brogniez), i. 97.

Œdema glottidis, i. 557.

—— in heart disease, i. 350.

Œdematous erysipelas, ii. 123.

Œsophagean canal, i. 184.

—— pillars, use of i. 138

Œsophagotomy, i. 164.
Œsophagus, i. 111.
—— action of the, in vomiting, i. 150.
—— description of the, i. 124.
—— dilatation of the lower end of, i. 152.
—— laceration of, i. 166.
—— sensibility of, i. 128.
—— ventriculosus, i. 164.
Œstrus bovinus, ii. 202.
—— case of, in a girl, ii. 209.
—— equi, i. 196.
—— hominis, ii. 207.
—— ovis, i. 685.
Ointments for local applications in skin diseases, ii. 117.
Olfactory bulbs, the, ii. 413.
—— nerves, the, 433.
Omasum, i. 131.
Omental hernia, i. 311.
Omentum, i. 236.
—— use of the, i. 287.
Open joint, i. 68.
Operation for artificial fistula of the stomach, i. 210.
Ophthalmia, periodic, ii. 372.
Ophthalmic nerve, the, ii. 437.
Optic commissure, the, ii. 413.
—— nerves, the, ii. 434.
Organic impurities, in water, i. 30.
Origin of fibres from a nervous centre, ii. 392.
Osmazone, i. 113.
Ossification of the heart, i. 372.
Osteo-dentine, i. 55.
Osteoporosis, i. 20.
Osteosarcoma, i. 72.
Outbreak of smallpox in Wiltshire in 1862, ii. 247.
Over-reach, ii. 593.
Ox, louse of, ii. 198.
—— tick, ii. 210.
Oxalate of urea, ii. 12.
—— of zinc, ointment in skin diseases, ii. 117.
Oxygen, i. 28.
Oxyurides, ii. 840.
Ozena, i. 543.
Pacinion bodies, ii. 393.
Palpation, application of, i. 535.

Palpitation of the heart, i. 851.
Pancreas, diseases of the, i. 320.
Pancreatic calculi, i. 320.
Parasitic, diseases of the liver, i. 318.
Pancreatic fistula, operation for, i. 248.
—— glands, i. 229.
—— juice, action of, on the food, i. 250.
—— juice, properties of, i. 249.
Pancreatine, i. 249.
Panniculus carnosus, ii. 67.
Papulous eruptions, ii. 126.
Paracentesis thoracis, i. 590.
Paralysis, ii. 496.
Paralytic staggers, i. 188.
Paraplegia, ii. 498.
Parasites, effects of, ii. 334.
—— in the œsophagus, i. 168.
—— in the stomach, i. 196.
Parasitic affections of the nose, i. 634.
—— disease of the lungs in sheep and calves, i. 637.
—— disease of lungs in calves and lambs, ii. 364.
—— diseases, ii. 333.
—— growths of the heart, i. 375.
Paratrimma erythema, ii. 121.
Parenchymatous tissue, i. 459.
Paronychia impetiginosa, ii. 139.
Parotid glands, the, i. 108.
—— of the dog, i. 109.
Parotitis, i. 121.
Paroxysm, i. 192.
" Parrot month," i. 11.
Parturition fever in ewes, ii. 318.
Parturient apoplexy in cows, ii. 313.
Passive hæmorrhage, i. 448.
Pathetic nerve, ii. 436.
Pathological anatomy of pleuræ, by M. St. Cyr, i. 583.
Pathology of the incisors, i. 78.
Peculiarities of sheep, ii. 87.
Pedal bones, disease of, ii. 580.
Peduncles of the cerebellum, ii. 417.
Pelvis of the kidney, ii. 7.
Pemphigus ii. 145.
Pentastoma, ii. 343.
—— tænioides in the sheep.
Pepsine, i. 214.

Peptone, i. 216.
Percussion, method of, i. 499.
—— sounds elicited by, i. 529.
Pericarditis, i. 386.
Pericardium, description of the, i. 322.
Periodic ophthalmia, ii. 372.
Peritoneum, i. 178.
Peritonitis, i. 285.
—— as a result of ruptured colon, i. 272.
Permanganates, as a test for the purity of water, i. 30.
Peritoneum, i. 239.
Permanganic acid, i. 30.
Perspiration, sensible, ii. 75.
Pervious foramen ovale, i. 363.
Peyer's glands, i. 230.
Pharyngeal branch of the pneumo-gastric nerve, ii. 444.
—— choking, i. 161.
—— polypi, i. 157.
—— symptoms of, i. 157.
—— tumours causing roaring, i. 624.
Pharynx, i. 122.
—— the, i. 454.
Phlebitis, i. 431.
Phlebolites, i. 437.
Phlegmonous tumour near the left ear, i. 83.
Phlyctenoid herpes, ii. 145.
Phosphate of lime in animal tissues, i. 33.
—— of magnesia in tissues i. 33.
Phosphatic calculi, i. 254.
—— composition of, i. 255.
—— manures, the value of, i. 34.
—— urine, ii. 24.
Phosphoric acid in tissues, i. 33.
Phosphorus in tissues, i. 43.
Phrenic hernia, i. 311.
—— nerves, i. 149.
Phrenitie, ii. 480.
—— and stomach staggers with delirium, contrasted, i. 189.
Phthiriasis equi, ii. 213.
Phthisis pulmonalis verminales, i. 637, ii. 364.
Physical nervous actions, ii. 387.
Pica, i. 184.

Pig, muscles of the, ii. 227.
Pigment in the skin, seat of the, ii. 67.
Pigmentary matters of urine, ii. 16.
Pigs, epizootic aphtha in, ii. 268.
Piles, i. 303.
Pituitary body, the, ii. 413.
Plastic, i. 37.
Plethora, i. 32.
Pleura, i. 316.
Pleuræ, the, i. 463.
Pleurisy, i. 581.
Pleuro-pneumonia of cattle, i. 592.
—— report of the Commission on, i. 597.
Pleximeter, i. 501.
Plexus, definition of a, ii. 392.
Pneumogastric nerve, ii. 441.
Pneumonia, i. 577.
Poisoning, symptoms of, i. 224.
—— by animal irritants, i. 242.
Polydipsia, ii. 31.
Polyuria, ii. 30.
Pons varolii, the, 412.
Portal vein, i. 248.
Position of nail-holes, ii. 574.
Posterior perforated space, the, ii. 413.
—— pyramids of the medulla, ii. 411.
—— repeller, Ewing's, i. 100.
Potassio, tartrate of antimony, i. 225.
Poultry lousiness in the horse, ii. 218.
Pourriture aigue, ii. 321.
Precordial space, i. 388.
Prehension, i. 6.
Preputial calculi, ii. 62.
Prevention and treatment of scabies, ii. 77.
—— of pleuro-pneumonia, rules for the, i. 610.
Pricks from nails, iv.
Probang, i. 163.
—— use of the, i. 163.
Prolapsus ani, i. 301.
—— linguæ, i. 23.
Properties of the urine, ii. 9.
Proportions of water found in the different solids and fluids of the body, i. 27.

Prostate gland, the, ii. 8.
Protein, i. 48, 214.
Proximate principles, i. 26.
Prurigo, ii. 127.
Psoriasis, ii. 114, 134.
Psychical nervous actions, ii. 388.
Ptyaline, i. 115.
Ptyalism, i. 118.
—— treatment of, i. 119.
" Puffing the glym," i. 68.
Pulex irritans, ii. 201.
Pulmonary apoplexy, i. 568.
—— artery, the, i. 466.
—— lobules, i. 459.
Pulp, the tooth, i. 51.
Pulsation, variations in, i. 342.
Pulse, places for ascertaining the, in the domestic animals, i. 340.
—— in disease, i. 344.
—— rate of the, in different animals, i. 336.
—— terms applied to the, i. 344.
Puncturing for tympanitis. i. 174.
Purposes of hair, ii. 98.
Purpura hæmorrhagica, ii. 151.
Purulent infection, i. 71.
Pus in the urine, ii. 29.
Pustula maligna, ii. 275.
Pustular eruptions, ii. 185.
—— eruptions of the heels in horses, ii. 189.
Pustule of cow-pox, ii. 287.
Pyæmia agnorum, ii. 326.
Pylorus, i. 134.
Pyramids of Ferrein, ii. 6.
—— of Malpighi, ii. 6.
Pyrecchymosis, ii. 151.

Quantity of blood in animals, i. 477.
Quilled suture, i. 178.
Quinsy of pig, i. 560.
Quittor, ii. 597.

Rabies, i. 157.
—— canine, ii. 451.
—— in the horse, ox, sheep, and pig, ii. 456.
Rabot odontriteur. i. 100.
Racemose glands, i. 108.
Râles, i. 522.

Ranula, i. 119.
" Rat tails," ii. 134.
Reaction of urine, ii. 22.
Rectal polypi, i. 302.
Recto-vaginal fistula, i. 299.
Rectum, i. 230.
Recurrent nerve, the, i. 457, ii. 443.
—— sensibility, ii. 395.
—— mange, ii. 133.
—— softening of the cord, ii. 498.
" Red Soldier," ii. 310.
—— water in cows, ii. 47.
—— in sheep, ii. 321.
Reflex action, ii. 386.
—— of the spinal cord, ii. 404.
—— on two sides simultaneously, ii. 405.
Regurgitation, i. 122.
—— action of the diaphragm and abdominal muscles in, i. 189.
Relative size of the itch insects, ii. 175.
Renal artery, aneurism of the, i. 422.
—— calculi, ii. 57.
Rennet, action of, on milk, i. 46.
Respiration in different animals, i. 449.
—— influence of the medulla oblongata on, ii. 421.
—— mechanism of, i. 467.
—— rate of, i. 473.
Respiratory apparatus in our domestic animals, i. 450.
Restiform bodies, the, ii. 411.
Rete mucosum, ii. 66.
Retention of urine, ii. 35.
Reticulum, i. 131.
Rhubarb, action of, on the urine, ii. 29.
Rhythm, i. 370.
Rinderpest, ii. 259.
Ringworm, ii. 147.
Roaring, i. 620.
Rodent, movement of jaws in, i. 49.
Rogne pied, ii. 547.
Roman bath, construction of, for cattle, ii. 98.
Roseola, ii. 126.
Rot in sheep, ii. 357.

Round worms, ii. 840.
Rubeola, ii. 226.
Rumen, action of the, on food, i. 136.
—— capacity of, i. 142.
—— movements of the, i. 136.
—— of camels, i. 131.
Ruminants, chewing in, i. 67.
—— stomachs of, i. 130.
Rumination, act of, i. 129.
—— time requisite for, i. 141.
Rupture of the œsophagus, symptoms of, i. 166.
Ruptured colon, i. 271.
—— rectum, i. 272.
—— stomach, i. 155, 270.
Ruptures of the heart and of the vessels in its vicinity, i. 358.
Russian cattle plague, ii. 259.

Saccharine diabetes, ii. 33.
Sacrum, i. 233.
Saddle galls, 121.
St Vitus' Dance, ii. 465.
Saliva, action of, on the food, i. 116.
—— chemical constitution of, i. 112.
—— epizootic aphtha, i. 119.
—— the, in mastication, i. 66.
—— quantity secreted, i. 111.
—— secretion of, how induced, i. 112.
—— the parotidean and submaxillary secretions contrasted, i. 114.
Salivary apparatus, diseases of, i. 118.
—— calculi, i. 120.
—— ducts, dilatations of, i. 119.
—— fistulæ, i. 120.
—— glands, the, i. 107.
—— glands, size of, i. 111.
—— secretion, the, i. 110.
Salving sheep, evils of, 178.
Sand-crack, ii. 587.
Sanguineous ascites, ii. 321.
Sarcoptes hippopodos, ii. 170.
Scab in the lower animals, ii. 155.
Scabies, ii. 155.
—— equi sarcoptica, ii. 165.
—— dermatodectica, ii. 167.

Scabies, symbiotica, ii. 170.
—— history of observations referring to, ii. 157.
—— ovis dermatodectica, ii. 171.
—— prevention and treatment of, ii. 177.
Scalesiasis, ii. 343.
Scarlet fever, ii. 228.
Scarlatina in a pony (by Mr Hunting), ii. 229.
Sclerostoma, ii. 342.
Sclerostoma syngamus, i. 638.
Scorbutic eruptions, ii. 151.
Scrofuloderma, ii. 223.
Scrofulous disease of lower jaw in horse, i. 71.
Scrotal hernia, i. 307.
Scudda, ii. 93.
Scutiform cartilage, i. 83.
Sebaceous glands, ii. 72.
—— secretion, ii. 76.
Secretions of the stomach, i. 205.
Secreting structures of horn, ii. 513.
Seedy toe, ii. 596.
Semilunar valves, i. 326.
Separation of the spinal cord from its connection with the brain, ii. 403.
Septum nasi, i. 111.
Shaving horses, ii. 96.
Shearing sheep, ii. 90.
Sheep bot., i. 634.
—— baths, ii. 90.
—— classification of breeds, ii. 86.
—— scab, ii. 171.
—— small-pox in, ii. 238.
Shoeing, ii. 520.
Sibilant râle, the, i. 524.
Signs of disease offered by the capillary circulation, i. 347.
Simple coryza, i. 539.
Singeing horses, ii. 97.
Skin, ii. 65.
—— and hairs, diseases of the, ii. 112.
—— divisions of the, ii. 66.
—— thickness of the, ii. 69.
—— vermin on the, ii. 193.
Sleepy staggers, ii. 489.
Small intestine, coats of the, i. 228.

Small intestines, divisions of the, i. 227.
Smallpox in sheep, ii. 238.
—— outbreak of, in Wiltshire, in 1862, ii. 247.
Sodium. sulpho-cyanide of, in the saliva, i. 89.
Softening of the brain, ii. 491.
Soft palate, i. 19, 452.
Solid excrements, composition of, i. 253.
Sore throat, i. 555.
Sounds of the heart, i. 332.
Sources of animal heat, i. 494.
Specific gravity of urine, ii. 21.
—— ophthalmia, ii. 372.
Sphenoidal sinuses, the, i. 453.
Spina ventosa, i. 72.
Spinal accessory nerve, the, ii. 446.
—— cord, action of, on the movements of the intestines. nutrition, and animal heat, ii. 406.
—— anatomy and physiology of the, ii. 397.
—— connected with the brain, ii. 403.
—— minute anatomy of the, ii. 399.
—— nerves, the, ii. 399.
Spinitis, ii. 498.
Spiroptera, i. 196 ; ii. 341.
—— strongylina, i. 196
—— sanguinolenta, 169.
Spleen, i. 144.
Splenic apoplexy, ii. 286.
Splenization, i. 569.
Spontaneous aneurism, i. 419.
Sporadic pleuro-pneumonia, i. 590.
Staggers, sleepy, ii. 489.
Starch in food, i. 40.
Steppe disease, i. 594.
Stercoral concretions, i. 254.
Stethoscope, i. 497.
Sthenic hæmaturia, ii. 45.
" Stiffness," the, i. 34.
Stimuli, ii. 386.
Stomach, abnormal deviations in size of the, i. 218.
—— bots, i. 198.
—— movements of the food in the, i. 205.

Stomach, muscular movements of the, i. 147.
—— mucous membrane of the, i. 206.
—— of carnivora, i. 145.
—— of the horse, i. 144.
—— of the pig, i. 144.
—— staggers, i. 178.
—— in the horse, i. 187.
—— in cattle, i. 179.
—— solvent function of the, i. 205.
Stomachs of ruminants, capacity of, i. 180.
Stomatanthrax hordeolum, ii. 308.
Strangulation of the intestines by pedunculated tumours, i. 278.
Strangury, ii. 84.
Strongyli, i. 196 ; ii. 342.
Strongylus armatus, i. 257.
—— armatus varietatis minoris in aneurism, i. 425.
—— cernuus, i. 259.
—— contortus, i. 196, 259.
—— of the pig, i. 688.
—— filaria in calves and lambs, i. 688, ii. 364.
—— micrurus of the calf, i. 638.
—— paradoxus in pigs, ii. 366.
—— radiatus, i. 258.
Structure of the œsophagus, i. 165.
Sturdy, ii. 350.
Stylo maxillaris, i. 65.
Subcrepitant râle, the, i. 526.
Sublingual glands, i. 109.
—— secretion, characters of, i. 115.
Submaxillary glands, i. 109.
Subzygomatic gland, i. 110.
Succussion, use of, in diagnosis, i. 536.
Sucking worms, ii. 343.
Sugar in the urine, ii. 28.
Superior maxillary division of the fifth pair of nerves, ii. 437.
Supernumerary molars, i. 82.
Suppression of urine, ii. 84.
Surfeit in a mare, by Mr Percivall, ii. 124.
Swallowing, the act of, i. 126.
Sweat glands, ii. 70.
Swine, louse of the, ii. 199.
Sylvius, fissure of, ii. 412.
Sympathetic system, ii. 396.

Symptoms of sturdy, variations in the, ii. 354.
—— of the Steppe disease, ii. 260.
Systematic remedies for skin diseases, ii. 116.

Table of breeds, &c. of sheep, ii. 88.
Tabular statement as to smallpox in sheep in Wiltshire, ii. 250.
Tæniæ, ii. 339.
—— cucumerina, i. 260.
—— denticulata, i. 259.
—— echinococcus, ii. 349.
—— expansa, i. 259.
—— mamillana, i. 258.
—— perfoliata, i. 258.
—— plicata, i. 258.
—— serrata, i. 260.
—— solium in man, ii. 343.
—— observed in the horse, i. 258.
Tapeworm in the sheep in Australia, i. 259.
Tapeworms, ii. 339.
—— whose cystic forms are as yet unknown, ii. 339.
Taurocholic acid, i. 241.
Teeth, action of, in mastication, i. 67.
—— composition of, i. 51.
—— in herbivora, i. 53.
—— instruments for operations on the, i. 99.
—— of carnivora and omnivora, i. 58.
—— operation on the, i. 96.
—— removal of, i. 82.
—— simple and compound, i. 49.
Tegumentary system, i. 84.
Temperament in relation to pulsation, i. 343.
Temperature, action of, on the colour of hair, ii. 82.
—— of simple baths, ii. 112.
—— of the blood, i. 475.
Temporal arch, i. 87.
Temporo-maxillary joint, i. 68.
—— open joints, treatment for, i. 70.
Terminations of nerve-fibres, ii. 392.
Tenacity of hair, ii. 81.
Tenesmus, i. 301.
Tetanus, i. 71; ii. 466.

Tetanus from clipping horses, ii. 96.
"Tetter," ii. 145.
Thermo therapeia, ii. 108.
Thick wind, i. 634.
The third ventricle, ii. 414.
Thiret, i. 27.
Thorax, divisions of the, for auscultation, i. 505.
Thorter ill, ii. 506.
Thrombus, i. 430.
Thrush, ii. 581.
Ticks, ii. 210.
Toe-pieces, ii. 575.
Tongue, action of, in mastication, i. 67.
Tonsils, i. 107.
Toothache, signs of, in lower animals, i. 91.
Tooth forceps, Mr Gowing's, i. 102.
Trachea, description of the, i. 457.
Tracheitis, i. 562.
Tracheotomy, i. 558.
Transmission of mange from horse to man, ii. 166.
Traumatic hæmaturia, ii. 43.
—— heart ruptures, i. 35.
—— peritonitis, i. 286.
Treads, ii. 595.
Treatment of mange, ii. 190.
Trematode worms, ii. 343
Trembling, ii. 506.
Trepanning, the frontal sinuses, i. 548.
Trichodectes equi, ii. 200.
Trichinæ, ii. 341.
Tricocephali, ii. 341.
Trichosonæ, ii. 341.
Trifacial nerve, ii. 436.
Trigeminal nerve, ii. 436.
Trismus, ii. 466.
Trituration, i. 66.
The trocar, i. 174.
Trommer's test for sugar in urine, ii. 29.
Tuber cinerium, ii. 413.
Tubular glands, the, i. 229.
—— glands of the skin, ii. 70.
Tubuli uriniferi, ii. 6.
Tumour on the upper jaw, i. 72.
Tumours in the cranium, ii. 493.
—— of the heart, i. 374.

Tunica vaginalis, i. 307.
Turbinated bones, the, i. 453.
—— bone, abscess of the, i. 552.
Turnsick, ii. 350.
Tympanitis, i. 171.
—— causes of, i. 171.
—— nature of, i. 172.
Typhus, i. 275.
—— contagiosus boum, ii. 259.
—— in the pig, ii. 311.

Ulceration.
—— of the intestines, i. 283.
Ultimate elements, i. 26.
Umbilical hernia, i. 228, 305.
Umbilicus, i. 305.
Uracus pervious in calves and foals, ii. 239.
Uræmia, ii. 35.
Urea, ii. 10.
Ureous diuresis, ii. 32.
Urethral calculi, ii. 59.
Urethrotomy, ii. 64.
Ureters, the, ii. 7.
Urethral calculi, ii. 60.
Uric acid, ii. 12.
Urinary calculi, ii. 55.
—— system, the, ii. 2.
Urine, ii. 8.
—— amount of, secreted by different animals, ii. 9.
—— in disease, ii. 20.
—— properties of, ii. 9.
Uriniferous tubes, ii. 6.
Urinometer, ii. 21.
Urogravimeter, ii. 21.
Urticaria, ii. 124.

Vaccination, origin of, ii. 233.
Vagus nerve, ii. 441.
Valves of the heart, i. 325.
Varicella boum, ii. 256.
Varieties of Purpura hæmorrhagica, ii. 151.
Variola ovina, symptoms of, ii. 251.
Variolæ vaccinæ, ii. 233.
Variolous fevers, ii. 232.
Varix, i. 434.
Vascular tumours of the heart, i. 383.
Vena azygos, rupture of the, i. 359.
—— porta, i. 238.

Venous hæmorrhage, i. 429.
—— obstructions, i. 432.
—— pulse, i. 346.
Ventilation of stables, ii. 111.
Ventral hernia, i. 309.
Ventricles of the heart, i. 325.
Vermicular contraction of the œsophagus, i. 128.
Vermin in the skin, ii. 193.
Vertigo, ii. 474.
—— abdominal, i. 186.
—— in heart disease, i. 350.
Vesical calculi, ii. 59.
Vesicular eruptions, ii. 129.
Vestigium foraminis ovalis, i. 326.
Veterinus œstrus, i. 199.
Vis a fronte, i. 335.
—— a tergo, i. 334.
Vocal cords, the, i. 455.
Volvulus, i. 278.
Vomica, i. 373.
Vomiting, the act of, i. 147.
—— action of the stomach in, i. 149.
—— conditions favourable to, i. 150.
—— from distention of the stomach and spasm of the duodenum, case of, i. 153.
—— in the horse, i. 151.
—— in ruminants, i. 156.
—— in stomach staggers, i. 154.
—— mechanical impediments to, in the horse, i. 151.
—— treatment of, i. 156.
—— why so rare in the horse, i. 149.

Warbles, ii. 202.
Water, amount of, in urine, ii. 9.
—— as an alimentary principle, i. 28.
Weight of the encephalon and spinal cord as compared with that of the body, ii. 410.
Weight of shoes, ii. 573.
Weight of the spinal cord in different animals, ii. 398.
White fibres of the brain, ii. 420.
"White scour," i. 296.
Windpipe, auscultation of the, i. 507.

Windpipe deformities in the, caus-
   ing roaring, i. 626.
Wolf's teeth, i. 82.
Wood evil, i. 291.
Wool, kinds of, ii. 94.
Wornils, ii. 202.

Worms, ii. 339.
Wounds of the foot, ii.

Xyphoid cartilage, i. 239.

Yelk, composition of, ii. 85.

Lightning Source UK Ltd.
Milton Keynes UK
UKHW010801080121
376670UK00003B/767